REAL TIME

REAL TIME

THE LOCATION OF TIME IN THE FUTURE AND PAST UNIVERSE AND DIMENSIONS

John C. Robles

To order additional copies of this book, contact:
Palibrio
1663 Liberty Drive
Suite 200
Bloomington, IN 47403
Toll Free from the U.S.A 877.407.5847
Toll Free from Mexico 01.800.288.2243
Toll Free from Spain 900.866.949
From other International locations +1.812.671.9757
Fax: 01.812.355.1576
orders@palibrio.com
756375

CONTENTS

REAL TIME

Temporary aspersion new theories of the universe time space.
The location of time in the future and past universe and dimensions.
Alternate dimensions and parallel universes.
Dimensions and time in the cosmic.
See how black holes could work by compressing stellar walls Like springs stretched back Of the deep universe of dimensions, parallel And no, how was it believed before.

CHAPTER 1

EF1

IC 1101 time reference A1 to IC A 64

TIME IC A1 a IC A 64 SPACE

ET A CARIANAE POINT A1 A ETA C A64 IN
THE REAL UNIVERSAL TIME EXPAND TIME

Time is relative, that means IT COULD BE LOCATED THERE
AND IT COULD BE OVERTHERE, IT COULD BE THIS OR IT
COULD BE THAT, FURTHER. On the other hand, we know
time in the course of our lives and we refer to the time
line of the centuries, years, days, hours, minutes, seconds
and microseconds but time is more complex and is not
just that and embraces a lot of things like the dimensional
division or sometimes we use that time denomination for
those cases of dimensional divisions and parallel universes
because that's how scientists of the 20th century named
it, but time that has gone by in a fraction of its self is not

just time but all that includes it and affect it, anything that has occur in this universe, and not only exists our courses in this universe also a lot of things and phenomenon that can be measure with different times and what we are going to expose here are dimensional time of parallel universes and of alternate dimensions to our universe as well as time that controls the past and the future and it's in the past and in the future where this times are located. When we finish studying this volume you will have a concept that will help you study better and scientists didn't explain it all, only those who studied physics understand it, however here we are going to explain it easily for everyone, also we are going to get into things they haven't go yet and we will understand those times that there could be and we will define dimensional times, dimensions and parallel universes about what are they and how they really exist since they are going to be very important in our lives and will affect a lot of things and transport technologies and sidereal travels as well as the dominion of time itself and its understanding when other scientists go deeper in its study in the future in a medium to a long term by studying those bases that the author designed and exposed and when all of them can be proven scientifically because until now they are only temporal theories to prove.

On the other hand we expose that those explanations about it today are explanations scientists gave us with the radio astronomy and optical astronomy with normal telescopes from that contemporary epoch from the 20's until the 90's and even in the actual epoch, and with those observations with infrared rays and the light that reach the earth showed the universe but only based in those radio and light waves that created the explanations of the big bang generally talking and they proved it scientifically, but there are many phenomenon, theories and deductions that came with those astronomic studies of the modern era; one of those studies from other astronomers and physics and writers is time and possible dimensions and in our case we realize and designed some very good theories that could be very useful in the future to science and it will stimulate students of tomorrow and it was achieved after 40 years of studies with books, deductions and comparisons, and the study of other physics scientists coming to the next deductions and studies in a synthesis of how it should be, after studying dozens of theories of several physic scientists and observant and theories of modern astronomers of the modern astrophysics; and in this volume the author expose his studies in the field of astrophysics and guides you to the light for your understanding.

The astronomers due to they couldn't study this to understand this things but maybe they deduced them slightly since they were just discovering the CELESTIAL VAULT OF THE UNIVERSE to explain everything about the big bang, what you are going read in this book I hope inspires future generations and understand what it took the author to study 40 years of his life since he was a kid with other scientists theories and astronomy books and science statistics theories. And all that effort was crystallized by explaining TIME or part of it, to explain what they still can't understand very well, but still they knew some already in the 50's and 60's and it should be there as they exposed their observations just in other ways, but observing with telescopes it cannot get to this conclusions but with the more objective observations of the celestial vault and isolated physic phenomenon of the universe and united theories and that's why they gave explanations about light effects and gravitational waves and orbits without really knowing what gravitational waves were or maybe some did know but most of all because they were looking that there all bodies were affected and the matter of the universe; even so the scientists of the last 90 years make use of devices and science irrefutable instruments like optical technology and the famous and incredible radio telescope and gave us the knowledge of THE CELESTIAL VAULT OF THE BIG BANG and has us all overwhelm, but they didn't had much time to explain in a bigger aspects of it of parallel dimensions and parallel universes and they only could explain what they could see and the observations of this celestial vault that they catalogued and exposed in hundred of astronomy books of the last 100 years; and because they did those scientific studies with empiric and scientific proof of what is up there they achieve to define the universe but there still a lot of things to clarify and to decipher those Facts of Universal Existence and they are just starting, for objectives from areas that they will catalogue with the reduced technology that we have for it and they just be able to catalogue the particles that make the universe, so the author will explain especially what is time they cannot see but they can see it in experiments and observations but they cannot explain it real good because another kind of observation was need it for the celestial vault just deductive and about comparison and observation of other sciences and objective discrimination for many files from the modern era of all the effects and maybe a other astronomers that reached the same conclusions of our author but he is the first one to show them clearly with an understanding that will help a lot of students about time and all possibilities times that there are in the universe and what they really are, and the most important thing about this work is THE GREAT CELESTIAL VAULT AND TIME AND WHERE ARE THEY and what is call real time and where is it.

EF2

AUTHOR THEORIES ABOUT PROBABLE UNIVERSE

Universe theories and space time theories, parallel universes and universal gravitational waves and universe texture what it is made of, maybe is space emptiness in expansion of hydrogen and expanded helium or just expanded gravitational waves and thousand of subatomic particles that drives to a net like a spider's web with holes from where it can be affected while is getting out from the attraction of a star and its local laws or it could be affected by been with extra factors and sometimes with the famous black matter that accelerated the universe but it hasn't been totally proved but we know it exists. And to see that place we will go imaginarily to know how spaceships would travel and the light itself and other objects and also how many particles were disturbed that didn't travel a lot of time to get here and they accelerate to get to the earth and be detected by our telescopes like the unexplainable flying plates, how did they got here and where do they came from? Who are they and how did they dominated the matter to get here? Are we so separated or not? Did they travel trough parallel universes? How the light of the stars got here traveling hundreds of years, did eluded space time? Well, we will have to go back to the space emptiness through 10 solar systems of distance to Alpha Centaur, just supposing and to expose space emptiness near to a dark area close to the earth astronomically talking, what is there and what is it? Is an empty space or there is a range of expanded particles net also a range of gravity streams and expanded gravity cobweb and from where we could go through other universes because maybe we are in the extensive net of those cobwebs of gravitational flow where it could be holes where we can get in if we had a device that can separate the nets so we can go through that stellar web and go to other places also affected with cobweb streams that mark other spaces inside our own gravity streams by going through the middle of cables that support the universe and time where sometimes objects fall down by their own inside and accelerate something or get lost in other universes and disappear from this one and falls free getting lost in other dimensions and only light can travel directly maybe or light evades time by going so fast and never leave from that ancient place inside those nets that are ruled by unexplainable times for us and weaves a complicated cobweb and that is the matter the universe is made of. And with this short explanation we dare to explain how it could be getting in one of those gravitational streams and cross to another one and get lost in other universes or travel through the total emptiness between that universe and the next one where there is no time and it could surprise us how many there are while getting through those strips from the gravity cobwebs and that´s how we really know time is weaved or maybe those gravitational strips are time and expand the universe, only that could bring those objects from that far; if everything were ruled by our local gravity measures and everything were light distortion, if not it would be like exposed agents and if it is like that it would take thousands of years to travel there and this way it could be easier and we just have to get into those cables that support the universe and we could do it with some device from the future that can separate them, with people and robots that will study it first and classified them and by doing that we are going to be able to travel inside those nets of infinite diversity and for sure the human race will dominate some day, and one of the things that maybe is that the universe fall down naturally or was affected by those nets by its own mass, maybe is like the observation where the universe accelerated because of the black matter, that it's been studied by men today and the same observation of particles that couldn't get here but they did but it was supposed they couldn't get here in a very long trip and other objects not identified and more, although other theories of physics science explains it somehow like black holes and others, all together create this hypothesis of the texture of the universe like a big mix sponge full of temporal holes and dimensional short cuts to get there and see the universe like a gruyere cheese full of holes and passages of space time and gravity streams with cracked textures in the inside, it can be expanded artificially and disappear or just get in if they open naturally, maybe. Taking into consideration the amount of sights of non identified flights in a year and peculiar particles that couldn't have survive and all the theories proved today and all the others as well as evidences and physics effects that have being studied like the black holes also other singularities, everything take us to think this and because we haven't study enough lineal time that rules the matter and the only explanations for many phenomenon of the universe like the sudden acceleration and where the matter was affected universally and other experiments that prove that's how it can be for its study as well as the gamma of universes that could be parallel to our lineal existence to this solar system and its gravitational center of attraction and that way consider there are thousands of nets from where we could get through artificially or naturally to other universes with other times or just by gravitational streams and all this at the same time in some universes that are ruled by its gravitational waves and make them exist, and they exist in a super universe of atoms, after all 90% of the atom is emptiness said the scientist that made the atomic bomb or magnetism, so maybe it has several parallel universes or hundred of them and that's what we have to study and know there are holes as big as our own universe, celestial vaults equally big and to travel to them we just have to elude those gravity streams that rules them and separates them in different space times and exist here in other times in this same space.

And for that we are going to study it when we grow as a civilization the holes in time space and can travel to them, to the times that contains entire universes in the oldest super universe, and maybe they pass this matter from black hole to another or other ways and in others more stables with a lot of life like there is in this universal time space, here where we exist in parallel with them maybe and at the same time they exist in those times not yet studied that rule the universe and that maybe are just times and in every time there could exist a universe and we could just go from times through times in seconds of space times and in each one parallel universes and that would be our existence, be in one of those times with all the known galaxies in expansion and in every second to see other time space with other particles and other parallel universes to this one and we just have to discover how to go through from one time to another and that way travel from one universe to another, but we know we have to evade gravitational waves that rule this universe so our particles have the freedom of other molecular accelerations of the expansion or open the atoms and at the same time get out of this universe and get into another one that only affects particles in another time with other gravitational waves generated in other all different universe to this one and a lot others in other times separated by the cobwebs of the universal creation of particles controlled by time space in its molecular acceleration ruled by gravitational waves generated by the mass itself of the planets, stars and galaxies and the expansion of the universe itself and others as well.

On the other hand, we know there could be gravity influences from the local suns and galaxies and the gravitational influences of the expansion of the universe that create different time spaces and even temporal differences that maybe to travel to the universe to certain places we'll have to adjust the time of arrival in the epochs that we leave, for example, if we leave the earth on may 1rst of 2016 and travel to other suns we could get there in the departure trip to a few years in the future of the terrestrial calendar or in a few years to the past since each star or galaxy sets the time and rules a different time and only when we get there we will prove that, also remember we have to leave this universe and travel through the vortex of it and the next universe or by artificial or natural dimensions where there is pure emptiness or we could crash with an asteroid; THAT'S WHY UNTIL WE CREATE ARTIIFCIAL DIMENSIONS WITH A SPACESHIP OR GO THROUGH NATURAL DIMENSIONS WITHOUT OBSTACLES WE WILL REALLY TRAVEL THROUGH THE UNIVERSE OR ANY METEOR WOULD END WITH THE JOURNEY OF MILLIONS THAT COULD BE IN THIS NORMAL UNIVERSE and by doing that we'll have to go back to this universe and by doing that we'll lose time or we'll get here in the future of the earth and that's why we have to adjust those measures or we'll be different chronometric times since maybe EVERY STAR

WILL HAVE ITS OWN TEMPORAL INFLUENCES AND EVERY GALAXY AND EVERY DISTANCE BETWEEN GALAXIES AND STARS WHERE EXPANSION WILL BE DIFFERENT AND LESS OF IT OR STRONGER.

REAL TIME

It's known that time is the amount of time that an electron spins around the atom and also it's known all of them are subatomic particles that could vary by having us anchored in this universe and time is the past and the future and time also is what it takes something to get from one place to another in the universe dimensions; in the course of an object or of a person or thing existing graduated in years, hours, seconds, microseconds and millionth of a second and being in a universe in expansion at incredible speeds because time is also the time it takes to travel through the cosmos all we see and all we are as matter as well as other rotation journeys and translation of planets, solar systems and galaxies that's why all those atoms that are part of them, but by observing phenomenon of different causes astronomers know there are other spaces in other times existing in this same space but in other time as well as other dimensions in other areas of other spaces which are defined by time where they fit and belong, that is why time is a lot of things but it reduces to simple formulas that scientists will study the next centuries.

The atoms of the visible universe in different times and other universes also in different times alternated here and from the future and the past also there and here that means from one side and from the other in a multi universe in expansion coexisting with other several universes immediately aside from it in microseconds of atomic difference that makes them fit with their atoms respectively that's why only the speeds we see in this provable optical universe are from this time, at least in the nearness of our solar system since there is evidence that every sun and every galaxy have different time independently how their photons or light aces get here or their radiation waves, ultraviolet and x rays and many others inclusive how their gravitational wave travel since they could be in differential and traveling through the universe and could get to the past if is not fixed by other universes or with techniques from the future to get to that time, for example, on June 10th of 2016 to get to suns near this one at 100 light years of distance maybe we´ll get there in earth time in years in the past of the earth or years in the future and we have to adjust that before that journey and if we go to Andromeda we´ll have to adjust even more maybe centuries or just decades, because time is ruled by atomic speeds that at the same time rule the gravitational waves and as it gets out of every near solar influence it changes maybe, but we only would know that when we get there to the biggest odyssey of human kind, that is just beginning, even so there will be the past and the future and there we will catalog only in that moment what it is and where is it, maybe there are only phases areas and coils of time in each area or we´ll have to strictly change of dimension or universe to get to the wanted date even to evade those distances and also to not to crash with a rock at a great speed thing that this author always thought, without dimensions it would be impossible to get there and come back or just impossible to get there.

------------------------------------OO------------------------------------

And so our author exposes in the subject of temporal astronomy

EF3

Time, as all subjects about the universe and space and specially time in its definitions and classification of what it is and how it really is not just limited to the time measurers of watches of hours and days, but making a careful study in its complexity and localization, something extraordinary for this epoch from the 21st century, we will take for reference the observation from recognized astronomers and from the theories already written but not so clear about it from such astronomers from the 30´s; and because of that we will expose some theories long for the modern science yet not written, very fresh and very easy to understand from this author that studied a lot of thing with a special directed information to understand it as a child and ended working as a geopolitics consultant, that information was given from his adoptive parents, he got into thoughts and studied 40 years of astrophysics and special documentaries as well, since that stimulated him to help in what he did and that way have a common purpose to help so that way he could get to those conclusions and develop this temporal theories and will be on discussion to clarify it scientifically some day in the future by other experts and that will awake unimaginable envies from paid out skeptical and even by deflected payments from counterfeiters some say, but at the end the truth will win and his work will be classified by experts scientists and also they will be stimulated to study those areas saving them a lot of time.

On the other hand, all theories of that kind are controversial at first, but because he studied the great ones of modern astronomy since they will be very real angles from where we are and what is time for real and where is it.

That is based in a lot of explanations for it and the discrimination of objectives took him that way and his own science investigations; so that way we will go understanding little by little and if is not exactly that way we dare to say that there wouldn't be a lot to think about it today including viable space travels and travels in time itself and natural phenomenon 100% wouldn't have explanations and that's why we wrote for you and your understanding hoping to stimulate tomorrows astrophysics science, with reserved rights for the author Juan Carlos Robles Guerra and you will find in books and articles as a reference and although they are for the future about a science that is just beginning, in time study it's needed to be clear about controversies about it for once and in this volume we will expose what time is and what other names has it and how is it really defined, also we will expose how and where is located according to studies of the exact science from the last 90 years and as reference the observations as well from astrophysics as atomic quantum physics and from the universe structures, its particles from 40 years of private studies of the author and over all cataloged of the only possibilities about this controversial theme today, also for the entertainment of some and at the same time in a few volumes, it will be fun to study time as well as its variants and even they will be always theories ready to be proven, they will be there, saved in pure physics books and some others of entertainment like the book "Escape to Century 25th" the author takes it in a entertain way for small and grown and he reserved it this way and the science theories in other ways, some are written at the end of this book and that way it could get to more readers and have fun at the same time and in the future it will be there for scientists and astronomers for its debate so other scientists can prove it; and today it is written in entertainment and pure modern science books like this one of infinite time or real time since they are the first written and articles about what really space time could be, about a same space but in a different time with a simple and practical explanation.

Reviewing studies since the 20's from renamed authors of the universe theme, we know the universe is expanding and somehow it will contract someday where it began, that means where all the matter came from and it's expanding but in not visible for the optical telescopes and infrared, ultraviolet and x rays radio telescopes, that's why we just can see a part of the universe in expansion and until a limit where expands, in a few words with the modern science we cannot see the limits of the universe but we study and deduce what is a celestial vault of thousand of million years of antiquity, 13 billion years more or less but even so we still speculate since modern science cannot prove it in some aspects, also there could have been one or two more universes before this one and it could be expanding in ancient celestial vaults of the universe, from the big bang and we know it because of the amount of elements in the matter as a reference since just one can fabricate from 3 to 6 elements and there are more than 100 and the time of the suns are 10 billion years, it means every sun cooks like 6 and our sun is from a third generation, that's why although there are suns of 2 billion years will just cook a few because the universe is 13 000 million years old and there wouldn't be time for all 100 elements and maybe there could have been 5to 25 universes before or more, since the theory they were formed in a mass from the beginning, is very fragile and that's why we won't ever know where are the limits but we know there are so and after that there is THE NOTHING that means no atoms of the universe texture and nothing can exist until the expansion gets there, that is to say the celestial vault is closed by the expansion of the universe itself or the previous universes, is the only thing that gives life to anything, and such thing it has been studied but not so much specially this theme like many themes of the parallel universes and thery have been ready to be studied but there hasn't been time enough from the modern astronomy to really clarify it, but in this beginnings of this exact science we will do it and keep doing it for hundreds and thousands of years and maybe millions of years as humanity.

And only is going to be more by studying and writing about it and its known and we know we are precursors of this theme of where is the universe located and where is time exactly located, if this one exist as part of some others and dimensions or it would be impossible to exist in other ways and it would be impossible to travel to the stars fast and elude time even if it is artificial or natural and would be a very divided universe where every previous theory to this one would be only inventions that can't be proven, but seeing all the evidences of it in natural phenomenon and possibilities of be or not be, we know that's how it is and for sure there are tens of natural dimensions and parallel universes to this one coexisting with this universe and we know they exist in a complexity never dreamed by humanity and will fascinate the studious for an eternity and to humanity since several observations and evidences point to it and we will resume it in this books of time and we will call them in different ways from this author with real time ---the location of time and escape from several centuries--- but the promotions and what will stimulate thousands in the future to study more and extend in this sense to space and time will take us to the knowledge of it, the author confess he thinks of this and defines as the beginning of the science of time since it would stimulate scientists and thousands of observers to go to the real path of studies of it from other angles that will save them decades and maybe more of time since the concentration of this books as all the books it will save them a lot of studies and scientific observation time by going in direct angles of studies in depth of thoughts and scientific proof that will take them to time itself of the universe and to the real definition of time.

On the other hand going back strictly to the theme, we star to expose that in the celestial vault all is concentrated in the study in the closest thing to us, that means in the expansion near our sensors and telescopes and that's why there are just a few writings and in our modern science and in this volume the author will take us to the limits of that possible universe with his 40 years of studies of astronomy and special documentaries and he will introduce us in the limits of the matter possible texture of the universe and what is made of and by consequence of space time defining it by the first time, how it could be placed every possible universe and every possible dimension and what they really are by discrimination of objectives and statistics of several studies of it and it will have to be proven by science of the near future and discuss it.

------------------------------------00------------------------------------

And so our author exposes in the subject of temporal astronomy

EF4

DIMENSIONALS DENOMINATIONS -- IC1101 A1 TO 64 --100 AND ETA CARINAE A1 TO 54 –100

These measures were designed and invented by the author to measure, to denominate and to define time and its dimensions for anyone who wants to use them if they like it while this theories are proven scientifically.

This means, in straight line to the galaxy IC 1101 there is time that belongs to a dimension called IC A1 and right next to it there are more like IC A2 IC A3 I CA4 and there is even IC B1 and IC E2, etc. and also there are the extra measures ETA CARINAE to define dimensions less important like A1 and ETA CARINAE A2 and also ETA CARINAE E2, etc. and those are invented denominations by our author to catalog dimensions and time from a reference point to another and that way understand it better and we are going to use them and they are the astronomic measures that could be used for several things, in this case for theories that has to be proven like the previous one and to study time and dimensions and we will use as reference IC A1 to IC 64 and IC of 64 that means in straight line from IC1101 galaxy to the earth or in straight line from the star ETA CARINAE to the earth, due to understand what are they it will be used that way, that means to separate 64 dimensions and 25 parallel universes in our theories and fragile parts with almost no matter of the universe from IC A1 to IC A64 to IC A 100 almost without atoms and after that there is the nothing, that means a space time without atoms and pure nothing.

Exposed in other episodes but even so the reference is important and it's in straight line to the galaxy IC 11 01 at a billion light years and here we will expose other measures designed by the author and are the most outstanding suns that are like the ETA CARINAE star in straight line to the earth and from there it will be use as a compass to measure or define dimensions and parallel universes, and in the next pages you will see the biggest reference astronomically talking and it will be the IC 11 01 and abbreviated it will be IC A1 to IC A 64-A 100 and that means at the same time abbreviated like that because we cannot measure from the beginning where the big bang began in the dark center in straight line to the end of the universe, but because we can't see from here not even with radio telescopes the center nor the end of the universe we have to use and denominate whit astronomic reference those lines and coordinates of IC 1101 and from ETA CARINAE A1in straight line to earth.

We expose that the dimension lines from A1 up to A64 and from universes will be from VORTEX A1 to VORTEX 25 and that way we will define and denominate dimensional time possible of great distances although we will want to really say is from the first point of the expansion towards the outskirts of the universe and in other occasions we will be talking and denominating just the next dimensions in an ascending numeration from less to more and from 1 to 64 and about vortex from 1 to 25 and about sub dimensions from A to Z to catalogue them as we keep getting into the time study so it can be easier for the students and the readers in general and for now on all those who want to use this measures designed by Juan

Carlos Robles G. and we will go showing them through our studies and next chapters to learn about time and dimensions as well as parallel universes.

BECAUSE OF THAT, TIME WILL BE

Here we expose that multi universe and time complexity and in the lines of time itself will always be variants that will define the multi universe capacities of store things because it always will be part of the celestial vault and they will be there whether we make them or if we open it like artificial parts or artificial doors of bubbles artificially made like the once it would use some ships to travel through the universe or for floating and other dimensional cavities and they were recorded in orbits atoms and electronic gravimetrical waves of the matter like in a hard disk and in different times since we are electrons in our material edge and also in the universe that means electricity but we don´t notice it that´s why we dare to say they´ll be there forever existing, yesterday, tomorrow, the after and the before recorded in microseconds to the infinite in different ways and will have measures designed by the author denominated IC A1 ETA CARINAE A1 to the other dimensional levels from A to F and from 1 to 64 levels and their variants explained in other chapters from his books and we will mention it later, also we will teach you simple graphics to understand the basics dimensions and parallel universes since we believe and know and deduce there are a lot of variants and a lot of universes also. Like all things we are physic matter of electrons, protons and neutrons and like the records of electromagnetism maybe they could be recorded in the celestial vault of the universe in expansion has we mentioned otherwise there couldn´t be time travels or the before and after nor dimensional artificial bubbles and the travels through space could last millenniums.

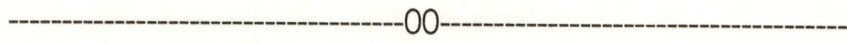

And so returning, to our theme of time

EF5

Alternate and altered lines of time do exist but a few times maybe artificially only and sometimes in other natural forms by clonation of material things by gravitational waves of the universe and maybe it could not be able to prove scientifically until we go there since the others depends on what beings exist in that universe and create them because any other possible way it just could be artificially created and depends on what kind of advanced technology there is in that place of the universe; although we could deduced it can only be made artificially or with several natural phenomenon in an alternate timeline where the same beings exist parallel to the other histories from a time, like if the United States exist in the civil war but without Abraham Lincoln being killed and stuff like that and when it cross the vortex to that time ETA CARINAE or IC A1 everything would be the same or if the attempt of Lincoln were aborted there would be a timeline that always will be there where it has been since happened once and because of that it created millions of years to the future in microseconds to the infinite of events instantly because we assert our existence it is unimaginable and we just have a glimpse of the knowledge of it, but going back to there could be another alternate temporal line existing in IC F B 2 where the president Lincoln is alive alternately in reference of IC F B from the future B parallel to IC A1 the will continue the same, alone unique for thousands of years.

On the other hand, take a good look at this is very important since there is strange evidence that we will explain in other occasion in the future, also there could exist very complex and diverse timelines and forms existing parallel alternating with our worlds in infinite alternate dimensions and which we have to travel there to prove it in a distant future, but sill today that would be fiction because we would have to accept a very complex and infinite multi universe of existential possibilities with an incredibly big diversity and we will explain it in other occasion so we won't complicate the explanation and so we will explain the basics of what time could be and where it could be situated if the next explanations of this author are expose and are the bases for this possible explanations of time, so new but they come from very basic scientific studies of several astrophysics from the last 90 years of modern astronomy and physic and they started in the 20's.

That's why we expose, alternate timelines and created when a attempt is aborted or stop building a building or changing something important in the alternate lines those that aren't IC A1 and IC A to 20 they could be destroyed due to it doesn't have previous real bases and are fragile and maybe the celestial vault claim its restoration or its finite, it means that dimension in any way that was occupied and created with a world or city inside it ceases to exist or just gets to a place and doesn't move on any more and unlike the others IC A1 to ICA20 more or less than being once and will reach to thousand years and even to one million years into the future instantly just by existing one time only, in every microsecond to the infinite this last type of time cannot be altered and if someone lives in it nothing changes if it goes to the past and kills it or eliminates it since it existed just one time first and there are the rules of ---JC.PF. It existed once; all the lines were created automatically thousand years ago and cannot be changed because they are interconnected automatically in every microsecond to the infinite on the contrary by being so vast the universe of travelers million years could change things every moment and people would cease to exist and would disappear in front of our eyes and also the buildings since something would always affect them in the past that's why only the alternate lines have to be taken care of or if they existed one time nothing changes them but others are created more and more and there could be 10 alternate timelines that we will study in other moment and that we will call "times of thousand" and they will be part of the studies of the possible time profundities, but for the moment we will focus to expose those closer times to ours and that they are alterations of the past or the future, alternated to our time, existing in parallel to ours from dimensions that only studies from tomorrow will define scientifically if only could be made artificially or they are created naturally in a ultra complex universe without limits, unimaginable, that's why today we explain it like this.

We know and we have to emphasize since we are talking about the matter existence in the universe and times or for the skeptical about the explanation of universal record in a hard disk of electrons with electron like a USB, really in this case the universe and its gravitational waves also we believe we could mention that between a millionth of a second and another there is a moment where there is nothing, if that once times were very strict, there wouldn't be atoms and there would exist the nothing like in other parts of the universe, like the real end of the expansion or of all the previous expansions, if there are a lot there we only would get to the NOTHING that means where there are no atoms or anything since we are just a CELESTIAL VAULT and that is it, with atoms in expansion of several things and diversities and if that time stops from one microseconds to another with maybe a neutrons machine or with something that could separate matter without damage, there could be a fissure or maybe there all ready is in every jump of a microsecond to another, if you are a believer of that once times and not of a multiple recording of the universe of several layers or thousands and time diversity.

------------------------------------00------------------------------------

And returning to the theme of dimentional astronomy

EF6

AUTHOR THEORIES OF ATOMIC FORCE FIELDS BORDERING PARALLEL UNIVERSES

In the possibilities and theories that we imagine and would have to be scientifically proven with preliminary tests, those possibilities exist that the author expose today; the things and objects of our universe from the famous big bang and it is an expanding universe towards different sides and they come from a not exactly clarify point from the center and an area of the universe that cannot be seen or reach with our telescopes.

In this universe all material objects, tangibles and at sight, according with this theory exist and they are materialized by the electrical charges of their atomic structures like physics know, and that is because they create force fields of different charges in their atomic structures and also they create gravitational fields that makes them repel from each other and therefore these objects from examples like water glasses to entire planets existing with the electrical charges of the orbits from its specific electrons and electronics create those force fields that repel the other objects, and to explain it better they delimit objects in their atomic frontiers and that's how limits of matter are formed, according to the amount of positive or negative charges and other neutrals of the atoms and because it is like that a charge limitation and create those force fields and have charge edges and forces that repels them, its created what we call or will call "the existential frontiers of the matter" like the electronic matter from a finger of the hand when touching a table, the two objects repel because of the electron limits in the edges of the atoms, but they are from this charges of this universe. So that way and the other atoms, when having other kind of charges in their electron orbits and when spinning at other speeds the electrons of those atoms or failing in other repel fields because of their negative and positive charges of their subatomic particles maybe since it has to be scientifically proven that those atoms create other parallel universes and exist near it or right there, but having other frontiers its objects exist atomically in other near areas from other dimensions and from a great celestial vault where exist many universes with different charges and different atomic bordering frontiers, and to go from one to another its needed some special force fields or induced gravity machines or which is invented in the future to change the spins electrons make to an atom or its speed or in other spins or speeds or positive or negative charges of them in their subatomic particles, once being there in those parallel universes what separates us from them is just charges differences and bordering frontiers of atomic charges and the atom so you can see in possibilities and just as a reference for normal people, the atom is 90% pure emptiness and the rest just electric charges and subatomic particles like some scientists exposed in the past and maybe you can sit in a chair and create emptiness and not stop until finding the universe of its charges if it were charged in other way in its atoms. So that is the explanation of why there are parallel universes next to ours immediately in the theory of the orange slices and only traveling to them it could be proven a 100%, but the secret by objectives discrimination it would be just what we previously exposed so it could be or not some different charges of atomic delimitation but several and several infinite or finite subatomic particles variants, but only that separates us from alternate dimensions and entire parallel universes in a same space but in different time, like theories from the past wanted to say but they couldn't find how to explain it and they knew because of experiments that was the only explanation for some phenomenon of relativity experiments and other theories and astronomic observations in the future at medium and long term, it can be seen the reaction and effects how light crashes with the objects and that way reflects every kind of atoms in an angle of photons frequency and consequently colors like that must be found or at least evidence of that kind of effects of subatomic particles reacting because of alternate dimensions of a parallel universe or dimensions of those, and it could be alternating and activating and changing possibilities of subatomic particles discovered after the positron in relation with its matter edges and the particles that result discovered or in relation with the effects obtained of the polarities and differentials of its positive or negative particles in relation with its protons nucleus so that way change the speed or value of its electrons changing its positives and negatives or just neutralizing somehow those polarities in an area where the atom changes of temporal value and fit in other universe and that's it and we could even do it with technology that would change our body atoms or those from a capsule to travel there and that way would just start to appear other things that cannot be seen in this dimension and just will be IC A1 to IC A2 or A3 or 4 or IC B1, and that's how it would be to travel to those dimensions with those changed atoms to other atomic speeds or differential of negative positive charges that fits there and we´ll just see how thing start to vanish and appear others and the object from this universe and this dimension IC A1we´ll see how starts to disappear and vanish in front of our eyes.

-----------------------------------00-------------------------------------

And returning to the theme of dimential astronomy

EF-6.2

THE BLADES OF TIME AND THE UNIVERSE LIKE THE MACHINE GUNS OF THE ANTIQUE AIRCRAFTS OF THE RED BARON SHOOTING AND NOT HITING THE AIRCRAFT BLADES.

In the explanations of the parallel universes we say they are in other atomic speed and that´s why they´re here or immediately here, but we cannot get there because we will have to change our atomic speed of perhaps the electrons by millionth of second to the atom or in microseconds to infinite; that's why we have to remember every time what some physicists in the past, that it is believed the atom is 98% pure emptiness or a 90% and if we don't fit in the electrons speed and our matter were in other speeds we would fall in an endless void if we were sitting in a chair, until we get to the universe that it is at our atomic speed of the electrons spins to the atom and that perhaps all its polarity values and speed to fit in our universe and the other one in its subatomic particles even all those no discovered yet, and polarities are created and very complex ones that create such spins and values. Like the machine of red baron shooting and not hiting the aircraft blades.

That's why we have to remember perhaps they are like blades spinning at a great speed and are in the road with each other but they don't crash, and they are there but in other spins and in other speeds and because matter is atoms at one kind of speed only since those atoms are in an endless blade spin spliced together, and at the same time with the differences of polarities of all of its subatomic particles that magnetically create those spins and complex existentialities and those that would be discovered through the years and centuries, but at the same time those atoms leave a record and existentiality in dimensions of cavities or vice versa that create the past and the future as those blades go through an intricate universe or gravitational waves universes or the stellar screen we have talked about in previous chapters. And those recordings really exist not just the parallel universes and dimensions or existed and will exist... and everything due to a splice and sophisticated insertion, like fan or aircrafts blades between them but without crashing, that leave dimensional recordings or failing for those skeptics as we say, create Thens of time backwards and forwards of those Thens, which means future pasts and for our author a recording in a giant hard disk that is the multiverse and the celestial vault of each universe with its dimensions only.

Created by atoms spinning like aircraft blades between them because of going in different speeds only and such spins create existentiality and the different matter of time and existentiality itself and at the same time the temporal recording or the Then of yesterday and tomorrow to which by other universes and dimensions we could jump and reach what it was there and what it would be in the future, through a universe or several strange ones without the laws we have here or simply without being ruled by the gravity waves and at the same time by the very different times from there the past future.

So, as going to that place of matter and blades, different, it would be a matter of understanding the movement of the blades of the multiverses life to go there.

And because of that we know and deduce that at least in theory that's how it could be, since such blades would be the walls of the dimensions or universes and of the rules of time that rule every universe.

And is very important to emphasize that perhaps because of that as coming out of them a little, at least in minimum differences of spins we could delay time or evade it, including in what we feel and aged since those spins of the electrons or blades without crashing gives the values to our atoms in their polarities and rule matter of thousands of forms in physics laws there are to understand and see centuries perhaps, but for something practical we know that an aircraft creating a difference in those blades when traveling the universe can evade time and see if we could make it to lunch time in a distant star at light years and depart hours before, since for them the time will not have passed only one hour, although in this universe to understand better as reference or in this temporal rigor rule they are a thousand light years, but also then we have to resolve another problem, to see if it can be done not just evade time rules but go back to the past just one hour later or ten according to the trip, when we come back to earth the trip would be in 4 or 20 hours only, something like that, since the spacecraft will go in an artificial space different in time that doesn't affect time itself from this universe…

CHAPTER 2

EF7

AUTHOR´S ONRANGE THEORY

to be proven scientificly
The orange slices theory of the universe.

And so, exposing the comprehension of the universe we dare to say a very small percentage has been studied and to understand the parallel universes and to even conceive traveling through the universe and the parallel universes that occupy the same space but at a different time but at the same time otherwise everything would be millions of years traveling and only androids could do it, so to understand it we will explain this theory of the ORANGE SLICES and it's a theory designed and made by Juan Carlos Robles where he exposes the universe is like a giant orange where we can travel in every slice all around the orange and every slice is an alternate universe and they cannot be seen one to another but only the entire orange, that means, in every inch there is a universe in the orange but by going through each slice it can go to the entire orange and even to the peel from the inside and not just to the slice because that's how the universe is and those slices could be even more complex spliced but we will explain it in another occasion since that is unlikely and not understandable, and so, with that orange slices model we will explain better the universe to our future astronomy students and studies of the universe in general of the parallel universes since we live in a multiple universe but all the alternate occupy the same space but in different time and so every slice is in a different time and the cavities are dimensions and those cavities would be interpreted as the smallest cavities of the orange so you can understand the universe that way only and not how we used to believed with the atoms with a lot of things that we didn't know that's why we will stay with the orange slices theory and we will just go from slice to slice with a special machine that will help us jump from our universes space time to another and we will go from one orange slice to another with inches of difference but once inside of each slice we could go to all the orange that means to all the limits of the orange that the peel delimits and that is the multi universe passing from one dimension to another that each slice has and further than the universe to another it would be able to travel to all the limits of the celestial vault independently in which slice you are in every slice would be a different universe and some in expansion and others just starting and maybe others will be extinct also there could be those in contraction returning from the expansion, but no matter how they are, they are in an existential slice where we could go to all the celestial vault that means to all the orange in the limits of the peel to the center of it where the expansion started with no obstacles and the slice wasn't small but an entire universe; we have to understand that's how it is and other explanation would be if we have an orange filled with grains of sand and every grain only fits with grains with the same size or with its same atomic structure and atomic speed or subatomic since the atom is 90% pure emptiness some scientists say it just like reference and they could fall down the vacuum if it were sitting in a chair and never stop until it finds a universe that fits its grains of sand and its speed and atomic structure because if it has many subatomic particles and a lot of orange slices and that's all, since the universe and the matter is not like we see it in this world but in other way, also the universe but it hasn't been explained because they're still to come thousands or even millions of years of evolution.

And so, the universe is like an orange that exploded from the inside out billion years ago and formed many orange slices with small spliced cavities that would be dimensions and every big slice would be another universe that occupy the same space but in other time, to explain this new theory to prove the multiple universe where we live in the sidereal cosmos... The end.

---------------------------------00-------------------------------------

EF7.2

GALAXIES IC 1101

Galaxy IC 1101 is the biggest galaxy of the universe and it is so big that to cross that galaxy you would need six millions of years traveling at light speed from one side to the other. It is the biggest concentration of stars of the universe which is located course to the expansion of the universe towards the outside. That means it passed through the celestial vault a long time ago about 1,000 million light years of the distance of the earth and it would be utilized as only reference. This is since in astronomy it is a unique point and a larger measure of matter accumulation of the universe of the largest and you would have to know that it is the biggest found until now. It is incredible gigantic and has a dimension of 60 water ways together which means 60 times 200 thousand light years and travelers may take a whole life inside it and there it would be the universe for many since we have doubts of what there is inside and that in its defect how it could be seen, on the inside of the universe with a lot of star stuff. There might be black holes of a hypermamivity similar to the big black area that describes the author in this book where he explains how at the beginning of the universe there are holes hyper massive and gigantic. One really big that maybe it has expulsed in some areas all the matter that we see in the universe besides the universal explosion just like that and simply only the biggest hole of the universe. Besides there could be others such as those of the end of the universe that the author explains that existed maybe at the end of the universe and they could be giant parades of matter that ends in one side and then even be lost in times. Strange where the laws of physics change enormously and multilayered and at the end where there is nothing but the strange singularity of the 100% nothing or zero % way to sustain the matter of the universe. That would be far from the end of the expansion and its masses and of the first universes and turned off stars, black holes and magnetic stars of galaxy type of hyper massive and that is why turning it to galaxy IC 1101 that we take as reference in this book for our astronomical measurements, called IC A1 and IC A2 to 64 as well as ETA CARINAE A1 until 64 that we used to measure the dimensions and our explanation of this galaxy that it is for you to know why we select it as a reference in our explanations since in our bellies there exists titanic singularities like a small universe of matter that maybe even in some parts it is independent of this universe. We don't really know how it behaves in the modern time that matter and we only see with our phenomena instruments like the near great attractor relatively of this local area of galaxies which attracts many galaxies to the chaos and to the end. That is why in the middle of the hyper massive galaxy IC 1101 there exists black holes hyper massive like the dolphins of the universe and the beginning that make the matter fall to unknown dimensions outside of this universe and time. And that it even communicates to this universe with other universes easily in thousands of singularities that there could be due to the effects of a lot of matter together in a sight and that maybe there is since the chaos until the unknown of the stars and for that same reason extraordinary civilizations and restless people to get to know us and lots of life. In other words, it is too bad that only when we denominate the matter it is when we could be able to go over there but maybe something here finishes like the technological stagnation or existential anonymous and that they come here and wonder and that they might have come to the neighborhood of the back of the expansion traveling maybe to the beginning of the universe since for them we are at the start of the universe and not like for us towards the end of the universe.

On the other hand, perhaps in their cavities of the nucleus there are titanic explosions of supernovas and of which black holes come out and similar singularities to the great attractor which is one of this galaxies and the destiny of the galaxies is drastic and very dangerous, different, chaotic and maybe we would have to give thanks to live in a galaxy that is stable literally speaking. Since there will only be chaos if a big explosion occurred and would put all galactic neighborhood in danger of the milk route when its nucleus explodes in form of galactic quasar and then fades like it happens at a determine time of years. Things that say have already happened several times when it overloads of eating suns in the middle of the galaxy in its hyper massive hole like the spiral galaxies and if one of those jets touch the earth or the turbulence it would be the end and it would burn everything and there would not be anything in thousands or millions of years more. The other part of danger of this stable galaxy would be far into the future when it crashes with the Andromeda galaxy in five thousand million years and by then we would be far. On the other hand, galaxy IC 1101 with 60 times de size of ours it will have 60 times more dangers that is why we don't have to envy anyone but for sure I could shelter civilizations much before than ours. Maybe they would dominate the universe and already dominate the matter millions of years and that is why there will be technology for even immortality and to prolong life hundreds of thousands of years or to have the most impossible suspension of aging and other wonders that here we don't have.

On the other hand, maybe on the galaxy IC 1101 there exist huge gravity seats that diverts anyone to cross the center of gravity without a modern spaceship for the best of the universe and for our rockets it would be impossible until the upcoming centuries and maybe one day someone will say "Hi" and gives us a shortcut or a ride and shows us some wonders that there are over there and maybe some indescribable dangers and wars because no civilizations that travel in universes after universes and to the end of the same.

On the other hand, we expose that there are many galaxies less large and like we said in other chapters, we don´t know from where it came all the matter that you might think it recycles form the universe or it simply was there at the beginning of times and the diverse universes created many matter through millennia and previous universes of the previously counted by the modern astronomy. It only starts to loom like our author indicates since we have to remember that to create an atom is incredible difficult and to create lots of atoms that there is in the universe is improbable and it might have taken lots of time and that is why many previous parallel universes that fabricated all those atoms and that at the same time created the incredible universe in expansion that we have barely discovered in the previous centuries and to understand and explain scientifically until the 20th century we proved what the bing bang was.

Things that we didn´t know before so from now on you don´t have to be scared when you see strange things like the ones explained on this volume of time, space and alternate dimensions and the parallel universes since it is part of the evolution of astronomy and not only of the logical deductions of previous astronomy but the consequences of the scientific tests of the observations and studies of the readers of astronomy like our author of the last documental and angles of subsequence of astronomical studies and in different themes like this one of time and its dimensions and everything that was exposed on this book that gives and proves a continuity to modern astronomy and the quantum physics of people that study, like our author, 40 years and of the studies in time and all the explanations that is what the continuity of modern astronomy and the profundity of the phenomena of its angles is and the physical object of time and its location in the universe.

The End

----------------------------------00------------------------------------

And thus seeing universal time

EF8 TIME SPLASH

On the other hand, we dare to say that in the beginning of the creation the explosion of the big bang by exploiting at different ways all the matter some celestial vaults were created and not just one since by exploiting there were time differentials to the expansive wave of that matter created also expansive waves of gravitational waves and they really are atomic expansive waves and of subatomic particles still not catalogued only in general today like gravity waves that created a celestial vault by going more and more through the distances to the expansion of the universe and we haven't prove if that happens in other already created vaults from previous expansions that find the quantity of elements that every expansion last since the elements are born in the stars like 3 or 5 elements, every existing sun with 10 billion years like ours and came from another one of 2 billion years and that one from another of one billion years and everything in the expansion that has from 13 to 15 billions of years old that started and because of that we dare to say that maybe there were like 10 universes before or maybe everything was formed in the great matter conglomeration like other scientists say and when everything exploded stayed like that but we'll leave that for another occasion since we are just going to talk about that expansion physically and we will get into it, so if there are 10 universes they all are in different temporal phases different spliced time differentials or atomic speeds of the matter edges and if it was only one explosion, by exploding time differential were creating or even we believe and the author agrees with the theory that there were 10 universes before and 10 celestial vaults before the big bang but even like that in each one diffracted the expansions in several gigantic serpentines and several dimensions or universes were created maybe in some universes that existed already like we exposed maybe 10 or in just one expansion and whatever they were, each one created different expansion waves from where gravitational waves diffracted from the expansion and the matter and which took different dimensional ways which created the several dimensional serpentines that every universe could have and maybe some different universes expanding slower in other times and others faster and maybe there is that in several universes and each one with several dimensions explained in the orange slice theory from the author Juan Carlos Robles and that's why the author expose there are maybe 64 livable dimensions with extra variants in an estimate more or less not to underrate the universe complexity by objectives discrimination and not to underestimate the human imagination because they could be less naturals and to many created artificially and that will be material for another text in the future.

So there could be very diverse dimensions each one from a to h maybe and maybe there could be others of 10 to 100 more where there could be almost nothing since they don't have atoms or there's only the nothing or dimensional vortex that takes you to the exits of the real total universe maybe, so that way the author catalogues such universes which they all are in the celestial vault and from now on we will talk about one celestial vault doe there are maybe 10 we´ll talk about the one that affect us first to understand the time we are in because it turns out that they were created like serpentines of spaces as they expanded to the nothing to the previous vaults taking with them stars and entire galaxies and they went out when it was expanding in the first phases of the universe and so they kept going more and more every time and splicing one to another since the expanding wave somehow it wasn´t uniform but very complex and that´s why dimensions were created and to tell in detail what are dimensions they are matter and gravitational waves included with suns and stars and galaxies that create gravity swirls like the swirl of the earth that has the moon trapped in an orbit and in which it moves away every time the swirl pushes it a little bit and it gets it closer when is tenuous and if it keeps going that way it could crash with the earth in 10 million years more and not before, but for that time we would have a method to stop that from happening and keep living here with controlled tides by the moon and with a stable balance in relation to our inclination with the sun. So we hope that explained this way it could be easier to understand what gravity is since they are particles waves charged with positive and negative differentials of the subatomic particles from the edges of the matter and even today they haven´t been all catalogued but we know that only that could be.

So when the universe exploded with all that matter created some celestial vaults with galaxies and stars that means suns and moons and in between those particles and that at the same time are GRAVITY in shapes of swirls that turn into complex forms and they start to get out and hiding in different times and existence speeds from the atoms material edges so that way created alternate dimensions and maybe even entire parallel universes where maybe the expansion has ended and maybe others go in the middle of ours since not everything was thrown evenly in the expansion or expansions creating, has we have said, the most discussed Parallel Universes in the modern era and at the same time they were cooling down. But if we keep going deeper in what they really are, there are discussions about if it is just gravity and gravity subatomic particles with multiple cavities, that means of subatomic particles waves what there is more in the universe and even some say hydrogen and expanded helium but the author expose us that maybe it´s the magnetism of all of that and with several parts together in a same space but in different times, explained in the orange slices theory of the author and the dimensional theories IC and ETA CARINAE.

And as you know and getting more into it on details in those gravitational waves of subatomic magnetism it shows that it is matter in different times what is solid and what is emptiness is just magnetism from that matter in expansion and which is divided in forms of different atomic speeds specially from its edges in its electrons frontiers and its positrons and all the particles that would be discover and isolated by tomorrows science after this ones and those there, but since we are talking about the edges of the matter we will concentrate in the interior of the atom that everything is edges and limits in the electrons even if they are affected by its interior and about the edge of the electrons magnetism that's why the only ones discovered and isolated in experiments today, the famous Positrons and the particles that show in our eras and are starting to be catalogued of that magnetism; but its known there are parallel universes as Einstein said and others that exist in the same space but in different time, only that could be, and they could be alone physically electronically inserted like in an orange fill with grains sand and every grain with its atomic insertion without looking to the other one only when passing the electronic magnetism barrier, already explained in the orange theory and to make it more clear, some scientists declared and others believe that the atom is 90% pure emptiness that's why there could be space for that if things were simple and linear in that case we'll be each one in each speed and that's it but the universe is not that simple because there is what is not as been explained until we go over there and from one dimension to another and the matter not only needs differentials but being at other atomic speeds in others exist the Nothing and areas that are here in different time but not only because of that but because that's how the universe is, with different areas not yet explained with different physics laws and different existential variants that only by going there would be cleared, maybe there are more variants and must be tens of them like empty existential charges of subatomic particles that create dimensions and unimaginable parallel universes.

On the other hand, we know that between dimensions and universes there would be the Nothing that means where there no expansion or matter and only where there is there could be matter and atoms and we think there is where the end of the universe could go and the end of the expansion but among us could exist and among the dimensions and the universes and the nothing is no atoms, but in some place perfectly empty we could open other dimensions of artificial particles with artificial bubbles and with machines from the future and at the same time timeless parts and in other times out of every universe and because of that out of any time and that way we could even travel faster to the stars since there would be other space, existential and speed laws or just go between the vortex, maybe that would give us the much desired time and distance aviation and even more like having our own private artificial universes and dimensions to live there in a distant future.

-------------------------------------OO-------------------------------------

And thus seeing universal time

EF9

That's why we expose, the universe was made from one or more universes in expansion so it was created due to as it was expanding very complex celestial vaults were created with shape of serpentines and with several layers that created several universes and that's why were the universes of the theory of the orange slices in different space times, explained in previous books and with a copy at the start of this book. That's why we dare to expose dimensions are defined as those theories explain in reference of space time and also they would delimit the universes and everything from a point of reference in this case is was the star Eta Carinae for near things and in this case we will talk about the depths of time, we will use the biggest reference in the universe that would be one of the biggest galaxies and at a great distance so we could measure more space and it is the galaxy IC 1101and is localized at one billion light years of distance and towards the expansion of the universe and it is huge and it's six million light years long and 60 milky ways and also is towards the expansion of the universe and which the author defines as reference of universe limitation in IC POINT A1 and IC POINT A2 to IC POINT to 64 and IC POINT B1 and to B 100IC b, c, d, e, f, g, h, i, etc and that's how special space time dimensions are going to be denominated and the universes space time in their possible terms and the vortex of those universes and from the dimensions that we will define as VX ETA IC 1to more, depending on the dimension, that we will further announce and explain and by now we will start with dimensions that arrange future and past times and we will call them FD from 1. 2. 3. 4. 5 and more and FD 1 2 3 and less until the infinite of each one to the future and from the past until the big bang begins.

And on the other hand we will show the images and illustrations in time graphics and the dimensions that could shelter those future times, subject to be checked by scientists of tomorrow and today, studied better and easier.

So we will speculate if matter and beings are been recorded and without a record as we started to live as matter or there is just then of past futures because that would be found out by the reflections of the readers and the scientists who can achieve to go there in the future, but it should be easier to study for sure and without referring to the study of light how physics and astronomers of our time slightly explained until now because this theme is just starting in the areas of quantum physics and modern astronomy and it hasn't been possible because as we said is just in its beginnings because this theme is the LIMIT of modern astronomy and modern physics from the last 90 years. That's why we are dedicated to create possibilities models and own studies of the universe without relying to the observation of the telescope and light only, nor the infrared radio waves and others that radio telescope get, but in a way of references and observations of those studies and others about the universe and 40 years of reflections and getting deep into it, that's why we'll say if we travel through time to the future there could be a few ways, and what is really the future and the past? Well we think is a part of the space where things are ahead and the past where they were and are related in some theories and scientific studies from several scientists and in other theories aren't and only exist one future and it's starting in this present although in the theories of this author is shown all the times exist and have to be studied the past as well as the future and maybe ours since we pretend to expose sometimes that our present DOESN'T RULE THE UNIVERSE TIME is just part of time itself and also we believe the present is just a time of something very complex, something like an everything already written or a little bit of everything, so that's why in the future in its universe cavities are all the microseconds to infinite of everyone and of everything so existing at the same time than the present or it would be impossible to travel through time and experiments and tests and even observations and rumors from beings that do it are many today and they're everywhere as well as speculations of real secret experiments of the govern, but what inspire more to study this theme to the scientists is the behavior of the matter in the universe as well as the light and other particles and atoms, also subatomic particles and objects of the universe, also all the rumors and what encourage more the author is the complexity of the universe itself and to explain the behavior of the matter that we will explain in another occasions because this time it could be observed like this and like other scientists in black holes and their limits like variation of light and space and the complexity that suffers by going through the universe. And on the other hand and it could be very contrasting with the universe complexities that it couldn't be like that since everything would be too far and it would be impossible to get from one point to another and most of the time everything turned out to be very practical when science advanced with travels from one place to another on earth since 300 years ago and those from today, also united to the avalanche of rumors that they are doing it and that there are stars with billions of years older than us, we know some being do it and in this volume will be expose all the possibilities that could be about it, by the wonderful discrimination of objectives that in the modern era would be the role models to follow for the studies of tomorrow, since reaching all those studies from others, the resume of all of that took the author to study and think and write in some cases like the orange slices theory the first articles about it, and it would take you to an easier study from more realistic points of reference of what could be space time in different variants for the understanding of the students of tomorrow and you.

That's why we know and expose, to go to the future we will have to go to a dimension necessarily and it would be reached when we see all of them and also we will have to go to a universe near to a vortex A2 or B2 or A of dimensions like IC F B2 or IC F B4 and from there we could go to dimensions where we could observe all the times as we had said, from a space out of everyone and from there we could advance according to many experiments and thousands of times to some places and established speeds in that point cero universe or half where we would be able to see all times like if we look from top to bottom from the complex orange and then we'll be back to dimension IC A1 but in the future and it would be ICF A1 or ICF B2.

On the other hand is has to be known that is like a computer's hard disk with everything recorded after all, all the universe is atoms in expansion and so are we with electrons orbits of limits of our matter and also the electromagnetic records of a disk, that's why we have to reach those future dimensions of diverse ways that are going to be exposed but first we will define them and we will go learning as we prove those logistic deduction from astrophysics statistics works.

Also we will leave it to free thoughts, what is or are some future "back then" in dimensions that as the universe is expanding to areas outside the beginning, those future dimensions that shelter those "back then", and in third term they will be just coordinates from other universe to get into this one when everything is over or when nothing has happen yet which we will define and denominate as FD A 1B CD 1000 000 and PD to 13 000 000 and it means dimensional future from one second to one million years and dimensional past of 1 million years although we know the past is 13 billion years we will just use the last million years and we will define with the next graphics of those possible times in this volume and future volumes of the author and some are in the illustrations.

----------------------------------00------------------------------------

And thus seeing universal time

EF10

REAL TIMELINE, IT HAS HAPPENED ONCE
AND A THOUSAND TIMES AS WELL.

And so, it could be that the continuity status of time it changes at least a few microseconds to infinite when is affected in the past since the line was already created and there are sometimes millions of years of that line in the future and it is very important has politics agents and parts that affect transcendentally to our civilization and our environment or society since everything happened once in fact just by existing and that came before nothing else, even if you go to the past from the future and undo that thing or kill that person it will keep existing in a timeline 80 to 90 years, because what it was done since it was born or since its transcendence was made, because they were already there once before any chrononaut or time traveler got there after once existed if it doesn't belong to that time and all the microseconds and millionths of a second as well as hour, days, years and decades they would be there once before you could affect them and you would have to kill all the microseconds in 80 years more or less or you'll have to use very sophisticated techniques to erase someone of the history and that really would be impossible, since everything happened once and even if is not recorded in cosmic parts and you are a believer of the theories of the "THEN" only and not that everything is recorded in the universe electronic as we explained before since you and all the things have electron edges and their frontiers even if they are complex animals they just are electrons form their atoms, like other things we magnetically record and the celestial vault is just gravity and complex forms of different subatomic particles, hundreds, maybe not yet discovered and with walls of other dimensions and other parallel universes has we previously exposed, everything in an eternal expansion, although, if just one "then" was there it created all the "then" of microseconds to the infinite before the intruders of other times arrived, that is why history cannot be easily alter and it could go to the yesterday but it would be already created those from tomorrow and so for 80 years and things a million years and maybe everything is written has some texts say even religious texts, and maybe you are more comfortable with the theory that everything is recorded in a hard disk of gravity expansion as we exposed and everything is recorded in a magnetic gravitational waves hard disk since we are just electron atoms of material frontier and the universe waves as well as hundred of forms even with vortex between alternate dimensions and universes and universal frontiers that are so hard that create black holes and pressure the matter and each layer and only one day with artificial dilution to soften the matter it could be by neutralizing it or polarize it from one side like charge it with more neutrons or protons and decompensate maybe real holes or vortex artificial worm holes would be made and it could be possible to go from one dimension to another, those are things that do not happen in the universe since black holes prove it and they just compress it or disintegrate it but it doesn't go to another universe, everything in the amazing odyssey of the universe expansion and its celestial vault in which is impossible to vary history in my previous theories or in other theories that I have used in my conventional writings, it could be done but alternating transcendental things a lot of years and maybe there are universal criminals that do it and in my theories like those from the book Escape to century 25th and Dr. 12.1 and exposes that maybe everything happened once and there is 80 years of microseconds of Then or it would change one Then by killing someone, but all the other Then were already created before the killer arrived and they would be the Then and not magnetic fields records but the Then; that's why we have to investigate and search if other timelines could be created where the real primary line always keeps going but other is created without the person, in complex worlds where there are many affected by chrononaut and there are several timelines in one odd time sequence that we haven't study and maybe they join depending on the importance of the person or object that changes, but maybe there is only one alternate time in dimensions that we denominate as FD, A1 to more or Dimensional future 1 to more or IC Point A1 to more, those are alternate dimensions that must be there, that could be possible but in the complexity of the universal creation they get capriciously altered or exist in strange ways that we'll study in detail someday and maybe they are infinite alternate dimensions and the primary keeps existing, it just need to be proven in the future with scientists and physics of tomorrow, even so, we have to warn it could be just a few days and then disappear and just have space for a few days or months an extra timeline or failing the chrononaut would have to vary a lot of things and that way really affect the primary line that would create a very high paradox in the future, even so we have to say that we have never seen buildings or people disappearing so is unlikely that those kinds of paradox exist so maybe just one extra timeline is created but we don't know how much it last and how it works in the looms of the universe and the alternate universes and its dimensions both physics and future time that means future past and is not known how could affect the chrononaut beings that could change those times so maybe is just not possible nor necessary, even so we will get into the possible records of the not real primary timelines and others artificial alternate, taking as magnetic tape the matter of the universe and as we expose in other previous books to this one, due to the edges of the matter in its atoms that is electrons with different speed that's why in other times fitting in other dimensions and universes.

We have study by objectives discrimination all the possible possibilities of this possible phenomenon of space time and the travels through time and is deduced there could be from different possible things like being protected by the Then already created to success once and cannot get there at least if we travel through all, it's known human beings that could travel in time from distant futures of thousand years, they could not change things at will since they could find strange gravitational forces of physics laws that we don't know, like rebound waves that perhaps push them away to other times or instantly clones them and their space ships or temporal capsules and it really everything is cloned when a timeline is created, in fact nature would do it or simply millions of years cannot be changed and only those that belong to the present can change the present and although it would be very selfish to think that way, we also know we should think in other alternatives of how alternate timelines are created and perhaps the universe acts naturally with shock waves that clone everything instantly when something change in the future or being altered and on one side everything is the same but in the other side the altered or changed by chrononaut mutates and copy in other alternate dimension with the entire world and maybe the universe of orange slices, it is incredibly a super multi universe of thousand different parallels and overall is formed by strange and complex unimaginable forces and also wonderful, after all is full of creation and life and infinite stars, perhaps also like that infinite physics sidereal laws and maybe because of that they can't come and talk to us if they are chrononaut of flying plates since they would damage materially or be affected if they don't belong to this space time or near this time and if they come from the distant future gravity waves, vortex and other radiations and strange waves would affect them and could crash with other times maybe and they could even get cloned physically and copied next to the dimension they were altering and they could end in other dimensions, and it's not known but perhaps vagaries of nature forces of the complexity of the universe take care of us.

On the other hand, there could be united to all of this, maybe ecological beings or others called gods, they could prevented to happen as they see into the future if someone alter something in the past, but we have to be aware there are physics laws that we don't have in this time and we don't know, maybe thousands of those kinds of laws, and that effect be one since by altering the past and not erase the future, the world where someone died, someone the chronotauts killed, creates an alternate future and the result of that would be a cloning of this world in other dimension since that point to the future instantly and effect affects them, also if there were beings protecting the times and there wouldn't be unalterable lines we would see many things disappearing in front of our eyes like natural forests by altering something in the past and more or there would be gods or beings that protected this planet since 50 million years ago, is less probable that much time, maybe half a million but not 50 million years, that time evolution would be abandoned and we would see those changes in it wasn't like that and on the other hand, if matter protects itself in this strange universe and if we have already thousands of millions years like this since it's only protected of changing and if alternate timelines are created altogether every time that time is alter by travelers in the past at least millions of years in the past and that's why we think times are protected by natural things and alternate timelines are created every time is changed and they keep existing every moment alternated to this time but also keep existing what already existed with no problem since if existed once it will be there for thousands of years in a sequence already made by been just once; and on the other hand in other planets or places of the universe it doesn't protect like that of nature or superior beings and maybe a few disappear or stay existentially suspended.

---------------------------------00---------------------------------------

And thus seeing universal time And so returning, to our theme of time

EF11

How many timelines could be created if we go to the past to make many changes frequently and how far would they go if they were really infinite.

In this special chapter we will study what could happen if nothing could be recorded anymore or if the tape only accepts some momentary records only.

Maybe nothing can be change because it would be known and would be punish and by creating a lot of timelines or seen the changes after thousands of years as evidence of the crime or change for other superior beings and only what it really helps it can be change and not what harm.

It could be that when is expanding, the speed clones things sometimes but something activates it and perhaps all things been related only maybe in an infallible way.

MAGNETIC RECORD OF THE MATTER

We know video records and sound are atoms recorded in other atoms because of their electronic charges of electrons in each atom orbits and we could be a record in a multi universe with thousand of cavities in different times and spaces of past future, that's why since that is what we are just atoms and the universe gravity atoms in expansion and that's it, that is why we know what there is or there could be a record of the matter in a giant hard disk which has existing the future and the past in a same time, if not it would be impossible to go to those Then or to those times because there has to be a cavity of Then and maybe it is a record depending on the universe structure in that era or area, since on the contrary anything that affect the past in 50 million years back like affecting species or plants would change everything, let's say somebody travel through time and kill someone 2000 year or 100 years ago if happened once stays recorded and they could create another record and other alternate Then, but what happened once could not be changed or we would see changes often since there wouldn't be time travels of other ways and considering the great quantity of civilization there is in thousands of trillion galaxies since we'll know it would affect us often by passing through here and they haven't at least in the last 50 million years or as we said we'll see many thing changing or disappearing often; and we are not what it was left of a whole so big, it couldn't be that way because of what we explained before and after in this volume, others would think we are the residue of what it was left but even so it's very difficult since we know the existing civilizations would be impossible they didn't change at all between them and not affecting our planet, like wars between those civilizations and changes of travels of diverse variant times of all the universe so they couldn't come here and change anything in 50 million years of the 3000 millions that evolution has, that's why we know it's more likely if everything is recorded in the universe in cavities that are explained in the orange slice theory and that way we know it's more probable that to exist the future and the past at the same time, it's unlikely that they could be separated they look like a record because they'll be separated in microseconds to infinite, but we know they are millionth of seconds to infinite and that's why it's almost a record; if you are the skeptical kind about the existence is not a record, the author is convinced it is a record since alternate futures are created and also it would be easier since video records are atoms and photons in a magnetic tape and the matter like planets and persons are electrons matter that means electricity and the gravitational waves are gravity waves in expansion that created the celestial vault, that's why we stay recorded in it in different times and that's how we exist in the past and in the future at the same time since we are born and in other observations we would be the matter that rules the times and for that we have to be exceptionally important and we are not at least in size in the universe, but even so we would be a record to the past and the future that we already create by growing in our lives and what we affect like buildings constructions, artworks or just our common life with kids of certain characteristics and with works that helped in some productivity and our present would be the creator of that record only but as we exist once all of our futures were formed automatically anyway and that is a record of any way we want it and in any way, don't you think?

Physics and astrophysics students of the future or rather there is a great possibility that we are a record in magnetic fields and maybe there are many alternate records since the beginning until the furthest future since everything is relative and in a real time.

------------------------------------OO------------------------------------

CHAPTER 3

EF12

TIME ESPACE CHANGE UNIVERSE TIME --BING BANG A1 ETACARIANE REFERENCE

Just a few thing in the planet are made to last like the black boxes made by man because they can support air, pressure and hundreds of things like sea water and are very strong and maybe gravitational waves are also very strong where existence itself of the universe is recorded, since existence itself that really exist like the moon, the earth and the universe which is confusing and not like everybody thinks and like a black box is indestructible since exist in base time and that time is impossible to vary since when it happened nobody never vary it and everything is predicted and also in that case all predicted or better said everything being interconnected, the events of our history and famous people is protected by being recorded in an inerasable timeline so humanity history itself can be protected too and nature itself;

fantastic theories to be proven like the existence of the reactor 6000 exposed in my books "Escape from 25th century" and other episodes, so, even like that, if the person is not important nobody can change the future by killing it or alter its existence since already happened and we can only travel to the second time that means Eta Carinae point AB2 or something like that and we will study it in other chapters in near future but what we are going to expose today is that is like a record in a hard disk of gravitational waves of the universe texture or the past already pasted of pasts and even if we change it, it already provoked thousands of events that were recorded in the future and written in the existing universes texture and existing times that could be several like a, b, c, d, e and 100 more with 1 to 50 each one or some 100 but from other times, measures that we will denominate Big Bang A1, Big Bang A2 and Big Bang B1 to Big Bang 65 perhaps and if we know the past but never our principal past since already happened once and we do see it but what happen is that the futures and the past Big Bang is the reference of the creation of the universe, is a simple law that means already happened and already created thousands of records of interconnected existences in the universe textures and maybe we would have to change a lot of things and many universes and a lot of big bang times to create a real paradox and interrupt the future by altering the past, otherwise we would see many changes constantly of the environment of our world, like if many things change in front of our eyes and disappear constantly since there are many civilizations that visit earth daily in space ships and they could have change something million years ago in the past if it wasn't that way. That's why what happens in the present creates a million years of events unchangeable and only the time studious of the future will clarify it and scientifically prove it, but we think that like every hard disk that works with electromagnetism, the universe is a gigantic hard disk where maybe times stay recorded for the skeptical but for simple minds that believe in stuff of space time more profound simply has happened and created futures divided in microseconds to infinite and one created the other, and even if we change the past, it already happened once it created that future and the future of immediate microsecond to that one and the immediate hour to that one and by laws of the universe are already created and are not alterable since it happened and maybe it could create an alternate future but the times Eta Carinae A point 1 can't change even if we get to that time or maybe we just reach the Eta Crinae A point 2 and not point 1 never or it just staid recorded and if you go to that time in a space capsule is already in history and you were part of that time at million years we believe only time could see that but even if its affected with a nuclear bomb it would alter were it didn't exploded that nuclear bomb since it was made and created with thousands of futures and millions of microseconds to infinite separated one from each other, times that staid recorded in the looms of time and history and that's why the time we are living in this universe of universes and parallel dimensions are unchangeable if its tried to do it in the past; and the author of this book thinks it has the end of the universe, there are no atoms in this time case it already happened and created a world of events and if even if you destroy the past the future of that remains recorded in the loom of the universe and you'll have to destroy every microsecond of the years that travel to the past, something impossible to do for the civilized worlds that could visit us and even like that the next future of this one maybe at one million years it would be there and only would kill the person in the past but not its future, that's why paradox doesn't exist according with the theories of existential possibilities that we will denominate times big bang A1 to more that is time measured since the explosion that created the expansion of the universe and its dimensions measures created from the expansion and existence and not the parallel universes or dimensions of dimensional times of itself like Eta carinae point A1 point A2 and so successively, but primary big bang times which we will detail later in other books.

---------------------------------00-------------------------------------

EF13

RJ LAWS

And has we said in this quantum physics studies. And that's how once was and the RG's laws of physics maybe someday they would say this, would dictate that objects and persons and any environment cannot be alter from the past, changing it and affect the future because a few minutes later we will find with what it was already there before we arrived and because of that the temporal record was already created or the Then before arrived and had happened and it can't since many things would have to be destroyed and in thousands of microseconds to the infinite since everything already happened once before it was changed before arrival and impossible to destroy if it was recorded in the celestial vault of the hard disk and dimensions and what everything is written by existing once, it can't change it would need to destroy entire dimensions.

And the chrononaut will arrive in any way since already happened and matter created all microseconds until the degradation of that timeline that happens every 3 million years maybe, but in the future the duration would be clarify.

And in respect of a person and an historical change even more or the same but in the terms of its physic situation it would be there since it has existed once and happened once and created that chain of the future before anything and before you arrived, that's why, to what it was there and not changing or otherwise thing would disappear every moment before your eyes and that's the prove it would disappear frequently in front of you.

It also could be that has the universe was expanding or other vagaries of nature or not yet study phenomenon create the cloning of timelines and exist alternately since the outpost clones things in dimensions and the speed of expansion sometimes clones things that sometimes are extraordinary, perhaps if we study other timelines if they were profoundly changed or they couldn't be, but something very special activates it and perhaps being related all things between them perhaps is one infallible way like if atoms call other atoms to accommodate because of the lack of millenniums it does affects millenniums in those lines or other less durable like a hose needs to homogenize water pressure to where is empty and another less good example is like airplanes transporting people and loads that feed people and transportation in other places and dimensional record cannot be deactivated since it's a hard disk that activates with reactions of extra electrons electric charges and when a transport of food or medicine to an area fail, people die and there's lack of electrons for the celestial vault and dimensional that perhaps claims it has an intelligent hard disk but those are very new theories like all this one's already speculative, even so, nobody changes almost nothing, perhaps if it travels through time and doesn't belong to that same time or if is not connected to that destiny and if it does creates a temporal crime that would be very significant only for tomorrow people since they would have and would see the other future and would change before their eyes if they had a machine for both there in millions of years or something like that, but what I'm trying to say is that nature demands it and would adjust and function in other ways or it couldn't be that way like an atomic bomb can't affect other dimensions in no ways, since dimensional texture is very strong and a black hole devours worlds and entire suns without entering to other dimensions, or what do you think?

And the celestial vaults claim electrons when things change and that's what activates the cloning since everything really is electricity only.

Even so, remember all this theories have to be proven and will give answers to hundreds of questions of scientists of today and from tomorrow, so it could be that if time can be changed and the future in a strange way or because was alter several times in a jump to the past that is not connected with the present of future and changing something like that maybe we'll create the undesired paradoxes that are nightmares in our science fiction movies and our dreams and if so we would be fragile and we have survive and perhaps we've always been protected by casualties or gods and superior beings maybe, and the universe is fragile and weak for life when time is dominated and the travels through it........ The end....

-----------------------------------00-------------------------------------

And so returning, to our theme of time

EF14

REAL TIME PART 2

Time and its variants of modern astronomy to scientifically be prove. Part 2

In modern astronomy there will be a lot of variants since there will be experts that have studied it on their own and because of that there will be more diverse options to enrich other studies when proved scientifically someday perhaps, especially if it's studied fervently and observed for many more years and especially today, in this book we will expose the studies of the celestial vault, that haven't been exposed in other book and places in the way we will and because we are pioneers in this theme we are glad to please you with this book to entertainment since it's what is next to see and learn y modern astronomy and perhaps it could get a scientific recognition, but theoretically is a very good theme to discuss and to someday prove it scientifically but that would take decades or centuries, because talking about time is just beginning.

It will always be about other areas studies that consequently bring curiosity and authors like ours that want to share with others his observations for entertainment first of all and not forget his deductions and to leave them embodied in books that will help other to understand variants and theories of time and space. So, has we delve in the subject of this book that talks about past time and future and how it could be and how it could be to others and of course if it is scientifically proved that a time like that exists and if time is qualified more complexly someday.

FIRST CHAPTER EF1

In this chapter the author will expose the theory that has to be scientifically proved in the future, but he takes us in an infallible statistics of what it could be and what was in the beginning of the universe and what is it based on the behavior of all the universe matter like supernovas, normal suns, galaxies in expansion, parallel universes and even black holes, we believe that in the beginning all matter should have behaved in the same way and that's why we expose this way things since everything is made of the same elements and materials in the universe and thanks to the science that study the light spectrum it was scientifically proved the fact that they are made the same way, that's why we think that in the beginning of the universe has the big bang occurred a black hole was created and it was massive, gigantic, the biggest of all, even bigger than those recently discovered in the middle of the galaxies and thousands of times bigger than those from the supernovas and from the beginning of the big bang that left in the beginning of the universe the expansion mass in its interior and it has to be in its basics principles like it happens with the other portions of matter in all the universe, like the rests of the supernovas when they explode and always expand enough and a part of the inside goes back and it has a return wave that creates a black hole for sure and with no doubt, but our instruments cannot measure those areas and distances since they do not have that kind of reach and it just clearly have 6 to 10 billion light years and we cannot see at that kind of distance as well as at the end of the universe through the expansion because of the same thing, but on the other hand we can deduce and calculate has we do in this chapter, that's why we dare to say, it had the same behavior of the black holes and supernovas and today in modern astronomy the massive galactic holes in the center of the galaxies, in the past we didn't know they existed and they are the pattern and that's why we deduce that was also how it was in the beginning and recounting what could have happened, that great big wave on reverse and what created, but before we count the following; we expose that at the beginning of the universe or of the first universe, because the author's idea is that there were many universes before this one, returning to the subject, the first universe perhaps is and it was a kind of poor atoms without big electrons activity or they were formed for the first time and for common people we could say it was something like a rare electric oils automatically concentrated that were condensed and were charging more and more every day and abbreviating until create a gigantic electro plasma that exploded once and after that 7 to 20 more times until it got to the first universe of known matter but more complex maybe with layers of different times, created basic suns and then they exploded and the internal part shrugged and after the wave of the first big expansion created return waves to the beginning of matter, time and space and exploded more and more every millennium and that way the famous big bang was created and hence the first black holes in different ways since they are in other times. On the other hand, perhaps with every universes explosion were created all the parallel relative times with those universes exploding many times and create a new one until the modern era that the last big bang created and at the same time created a wave back for sure but we can't see and maybe it has in its center a titanic black hole that devours everything in its way and perhaps it will explode again has the big bang theory says, that's why we know the fusion phenomenon occurred in the suns of those universes, but maybe there were universes capable of creating infinite matter because sometimes we wonder from where comes so many matter that doesn't ends and from one explosion only we don't believe it and matter with 117 molecular chemical elements and because we are in dippers of our civilization considering the number of years it would take life to end only in our sun, they are 5000 billion years let's not just say only in our galaxy or universe since we need to go to the end and to the beginning to understand better our universe and know who we are, because our modern telescopes can't reach dose distances and recent studies explain it, not even radio astronomy that has better range than optical astronomy can reach the beginning nor the end and there are only models of what might have been, that's why we say alluding to the last radio astronomy informs where there are images taken by orbital telescopes and probes from remote places with shadows of ideals planets and with other recent telescopes where they show going through this area millions of galaxies towards the expansion like walls of gigantic distances downward and upward of this galaxies area of incredible proportions traveling through the universe towards the expansion and where it looks like walls parallels to others and they create a mosaic where you can see infinite abysses between them as if there were a galaxies factory somewhere and they come in series, that is, it doesn't look like an explosion of expansion what transformed them but like a machinery in series that throw them to an area of the universe and they go without stopping infinitely filling walls and wall of galaxies, something amazing, and so we think is not just an expansion because of an explosion there has to be more, maybe they came from other universes or from vortexes but it would be precipitate to talk about that, it could be an optical joke of astronomy but sometimes they don't look like they came out from an explosion but from a factory in galaxies series, then we ask ourselves what is out there, is there is not a black hole but a white hole that generate them perhaps or a gigantic spinning sun that throws them endlessly and to other times and spaces and traps matter in its guts, or what is out there since is too much matter and maybe the explosion was bigger than what we thought and it was an hipper universe, but even like that why can't galaxies stop coming out from that beginning, so we wonder what is out there, we want to see what is a wall of galaxies and why there is no effects or there are no images of a hipper massive black hole because is so far away.

We know it has be too far away so here galaxies can go and go like made in series and the beginning of the explosion is not visible nor the great black hole there should be, also if it were a normal explosion of the universe that we are used to see because everything works like that mostly, in about 6000 millions of light years through the expansion and the beginning, so with the ranges of our measuring instruments we know and calculate there wouldn't be nothing more than a titanic explosion norms like the supernova and it would just throw matter and would have an end and a determinate trajectory and is like an endless tapestry of galaxies passing like an advertisements parade; what is that and what's there over there, it cannot be just something that the big bang dictates according to our author, there has to be something more and we can see the advance of the expansion only since modern astronomy has helped us see an endless galaxies and a tapestry that confuses and makes the skeptics think that and they always want proves and the prove the author gives us are the comparison of the rules of the near universe and of what could be over there, because if it was a normal explosion as the big bang dictates there would be an arch toward outside and not so many matter distributed that way while passing that area and maybe it really is that way but with much bigger source for the explosion, so big that we can't see it here only its effects and they are only different lights, and they are millions of galaxies going through almost without arch but like a strip by this area in the most recent modern observations and it must have been more than just a simple explosion that expands matter, and is not that far like five times our range maybe and it looks like we explained before, and that's why we know if it was a matter explosion only of normal agglutination it would be with more arch to the sides but even like that is too much matter.

2

We must clarify there has to be a hipper massive black hole, bigger than any other in the universe, fascinating in its center and tests us to discuss those possibilities and how it would be with its events horizon, everything a hole takes, so not even modern astronomy talks about it since no instruments reach that place and the explosion must be impressive and that black hole hipper super massive with maybe thousands of millions light years and its effect would be perhaps distribute matter differently, incredible things like those from normal black holes and maybe it might be a strange quasar that expels matter and it would be that way until thousands of millions times further since matter would get there but not the tapestry of galaxies, even so, will have to consider what is it and if it is another kind of black hole that eats everything or that expels everything and it could be a big black hole from another universe that perhaps broke the walls and webs of the universes and all that matter comes from other universe and it doesn't explode again like the big bang dictates, but one day science will clarify it.

-----------------------------------00-------------------------------------

And so returning, to our theme of time

EF15

On the other hand, many will ask, maybe, although in astronomy is not that way since it is pure science, who was first, well the first beings were from there for sure and before there was nothing, something like the nothing, for what no one has an answer only superior beings or gods maybe, but one day it will be clarified if ambitious humans let it, that's why we are dedicated to what science says about matter only as well as everything we see, know and observe, just like that and no other way could have been.

And returning to the theme of the beginning and how perhaps was achieved and what happened to it since we know it could be something exceptionally loud, but above all it would be like watching explode millions of galaxies at the same time and not a sun from one galaxy, but there are doubts about how we explained it, maybe matter was thrown in another way as we exposed to see and review and go deeper into that later.

Today we will talk about how it could be that black hole; we think it would be something terrifying, imagine the supermassive holes of a galaxy since they are a galaxy and its matter only but that black hole perhaps is different, perhaps is something that in spite of we cannot see it since it's beyond our reach and it's impossible to see we can only imagine it, is it something that eats entire galaxies every hour if there were galaxies there and it created parallel universes and parallel dimensions and I think also the celestial vault and all the universe we see today and the universes that were before maybe since something like that could create a universe so big, which by science and several theories, effects and laboratory tests and we know it created spaces in different times and because of that scientists created wonderful theories like the one of relativity and others, and at the same time stimulated others to think in other theories of parallel worlds and antimatter and today we will expose in this volume what the author created and studied in 40 years of study of astronomy and discrimination science and analytics, because when the instruments don't dominate something it has to be reach with mathematical calculations and by the way modern astronomy has forgotten a little bit since it has been busy with the optical part and radio astronomy and it there has been a lack of this calculations and like this deductions that hasn't been explained and other aspects and in this volume we will give you some of that that it will amaze and like and we will explain how the universes could be and how they are and how they exist in the same space but in different time thanks to those explosions in the beginning of the universe that created them several times according to the elements statistics that constitute the universe and the modern physics laws. We will show you the parallel universes and the author theories of the orange slices that the author invented for you and that will be scientifically proven someday and today we will speculate with calculations and we will define how dimensions could be and we'll even catalog them for you so you can study better time and space since we believe every explosion created a time perhaps with different dimensions and created incredible alternated celestial vaults in the universe and thousands of millions of galaxies traveling through it and we are in the great universal expansion of the known universe....

EF16

SPACETIME

MODERN ASTRONOMY REAL TIME 2

So the end of the universe can be made of several layers of old universes before this one and have as we exposed from 500 000 million to 800 000 millions of years maybe since there could be one or several celestial vaults of several previous universes to this one and in that end did not affect the last explosions if they are in this time or failing that some universes and its end, as we said they could be in other times since in every explosion there could have done some different primary universes in many times and just a few in this space time in its astronomic and electronic speed since atoms could combine with differences of atomic speeds in its electrons edges and an example could be what we have exposed about the parallel universes even so everyone would have parallel universes recently formed or at the same time with its explosion or from those who formed them, perhaps some universes were formed not from an explosion but by shooting matter only or spin and also every parallel universe would have different dimensions, several cavities and celestial vaults as we previously exposed and because of that the end of the universe is a great storage of waste perhaps with previous universes edges and the expansion of this one at the end with galaxies already off and thousands of almost getting off and a place incredibly dark and a really dangerous magnetic edge of black holes and white dwarfs and what was left of infinite old creations and as well as infinitely faraway maybe like 10 times the remoteness of the known universes considering what a stable star lives like ours about 10 000 million years more or less and to less stable stars of 2 to 4 thousand millions of years old and the expulsion used as reference everything will lead to this universe to almost 100 000 million of years through the expansion with all its lit matter being turned off and now dark matter or bright matter like suns and black holes and there when everything is off they would be magnetars and black galaxies that would eat everything on its way, maybe there are areas where only gravitational chaos rules and dimensional confusion and there is nothing more than gravity and going through there it would be terrible in this space time and maybe to go through another space time there would be traps in other dimensions that would devour and wrap in a nightmare to get out of there to those travelers of tomorrow and even so some day they would get out and see an advanced of a gigantic space emptiness almost static at its end like a quarter of the path of what it was of the turned off matter that means of the ten times of this universe and the last two parts or four turned off there would be a quarter part of everything and also it will take us to get there like some thousands of millions light years further to the edge or vortex of all universes and the space emptiness at the edge of push and there only emptiness maybe and then finally if it were the last expulsion from the last or longest universe from this time, there would be the denominated almost incredible THE ATOMIC NOTHING the absolute cero matter not even space emptiness, no atoms, no expanded helium nor hydrogen atoms, the real nothing, the nothing of the creation perhaps, where if matter gets in it would disintegrate since there wouldn't be nothing to keep it integrated at least in time and space and it will change and maybe there would be like a barrier of the nothing and something has to be done to live there that means it has to be created a celestial vault there of the expansion push when it gets there or ours artificial first so we could live there and even if the modern astronomers haven't talked a lot or nothing about this since we know they reject this explanations and they really know this concepts or for the lack of time or radio and light waves observations to prove and well it's impossible to get some lectures from there, but doing a statistics deductions of what there could be after the visible universe, this would be only the scene by objective discrimination from modern astrophysics that in modern era it will be used as there are more and more studies and information of different academic subjects or failing that just studding more and deepening in astronomy and make a kind of thesis of this kind of areas in astronomy, not seen, and the author gives us in this pages his studies of 40 years of studding the universe for you, hoping you like it and stimulate you. So there in those places that are characterized in the lines of this book, we expose you what is there and what must be there, the yes and the no of matter and all the secrets of the universe since they would be there to be discovered some day, like what remained of them and before the end there would be horrible tornados of space times and there would be endless strict present and endless ends and life without time or beginning and an end and a variety of thousands of strange things and incredible whims of nature and a place where diversity rules and life is more intelligent than what we have here in the golden era with shining of the stars and the middle start of the universe and life since being thousands of millions years older than the one we have here, it would be superior and will host all the secrets of the universe, maybe worlds that ceased to exist centuries and millenniums and thousands of millenniums of millions years ago and there was only left copies and machines only until there are left only electro plasma beings and of dark matter and of android synthetic parts and very advanced and some destroyed by the darkness and for sure you have imagine other newest universes in their middle beginnings like this one or its middle future present and there would be what no one could imagine but that would be the end of the extraordinary expansion that created all endless life.

EF16.2

TEMPORARY RECORDING 2

We have talked about time is a record of the "then" or gravitational rays and atoms in other atoms and everything has been recorded there in space in parallel dimensions and universes in the then and atoms and how everything is made of atoms and the atoms are cores of protons and neutrons and with electrons orbits and at the same time sub particles acting that maintain them in a stability that even scares since there is too much and very good organized and we wonder where it came from so much matter, so perfect like the atoms in a perfect balance and hard so hard that not even the atomic bombs can break and perhaps only the black holes can bend and compress and disintegrate, but the walls of the stellar screen of the celestial vault created by other atoms, is not defeated for sure and also they create those black holes maybe because it haven't been proved, but our author affirms that's the only thing can be, invincible walls if they are not given the right angle perhaps neutralizing its fireas in a subatomic way and the black holes are spring walls imprisoning matter in other ways, enter in its walls or perhaps just change the atoms; it's known the atoms were made by the electro plasmas from the center of the universe and the stars at incredible temperatures of nuclear electro plasma fusion and just that, and every star like ours in 10 000 million years of existence creates only like 4 atoms that throws to the universe as its external layers explode and makes a nova or supernova.

And from there on, modern science doesn't know what and how so many atoms were produced in the universe, that bunch of atoms are really unimaginable endless and incredibly big.

On the other hand, we will be dedicated to explain what we were exposing about the "then" the past and future and temporary recordings that are made of atoms only, atoms of the same value nailed in other times and atoms with other values nailed in times like ours, so it's known an atom is electron, proton and neutron and their sub particles since until they are all catalogued in a few years, there would be enough data to know better and today there are already antiproton and antineutron and 30 more but there would be like 200 to discover and we know it's when we change the atoms environment although from there will come many discoveries until they are catalogued everything would be known and that will take a few more years and if perhaps someday manipulating those subatomic particles would be the discovery of the other atoms and consequently the other dimensions and parallel universes and meanwhile we know they are there otherwise it would be impossible to travel the universe without crashing a floating rock, right? Specially at light speeds.

That's why understanding this also you have to know the other universes would be in other times and atomic composition accommodated in a way it would be possible to be in the same space but in different time, which means, near here at a few inches far or miles only but topped because they have other temporal electronic values or values of dimensional times, that our author barely catalogued with the measures ICA1 and Eta carinae A1 and AB to 64 and ICA1 and ICA2 to 64 and that way to give us references of what they are and how they should be exposed in the realities of the celestial vault and not in explanations without a good argument.

Returning to the dimensions and the recordings of all object of atoms in this universe in times or in other then or better when science take us to travel through time we will know many more universes so vast that perhaps our is the 10 percent of all or a maximum of 25 percent and it's big believe me, it even has other unimaginable times and futures perhaps in itself or perhaps this one is the vastest so we won't underestimate the hipper giant universe where we live.

That's why returning to our theme that exposes how an object is recorded in a then or a temporal recording either thens or space magnetic fields recordings that really are hydrogen and helium expanded at miles and other joint gravity fields from the expansive wave of the big bang great explosion and other universes fields previous to this one, like 10 perhaps everyone with its extra dimensions from 10 to 64 perhaps and were created when the universe exploded at the beginning and create the celestial vault of the big bang and be with different subsequences that at the same time created diverse extra dimensions, like those from the future, the past and the present, perfect future, perfect present and perfect past and the then of time, cavities of every microsecond to infinite are created and exist, for the skeptical time instants that we call "then" from the yesterday and tomorrows and for the logicians and more of quantum minds, the universe temporal recordings whatever they are since one seems like one thing and the other seems like other thing, but at the end they are almost the same thing since they would be thens of millionth of a second to infinite and not just microseconds, that's why going too close to each other in a record like the recording magnetic tapes and the voices and the light our atoms of our molecular structure that at the same time atoms nailed in this universe and everything around us like air, land, water, the planets and the stars as well as the galaxies even if they change a little perhaps going there.

On the other hand, that is time of the future and of tomorrow which means IC-F-A1 to the million and IC AP to the million that means from here to the past or to the future, dimensions that we will detail and we have been explaining along this book, what there should be if it wouldn't exist the phenomenon called TIME and wouldn't travel through the universe and stars? With Tomorrow's people---

Returning to the temporal recording, we have created rules that perhaps someday they'll become laws and we abbreviate them with syllables IC and ETA and F and P like the rules jr and jc and are the next, if you go to the past and burn a tree in a few seconds it would be back there again and you could keep burning it but it would keep returning since what it created it already was the first time and created all the thens until the end of thousands of millions years otherwise thing would disappear every moment, even so there are variants of explanations that someday they will proved, but could have those measures and deductions many years since it would be that time is intertwined and something is keeping it and for it would be the same, simpler measures and to the FM years or the FA one million and you could create other time perhaps by burning that tree and other future but in other dimensions but in our primary dimensions ICA1 it will be there otherwise you could see things changing every day in front of your eyes and you could see people disappearing in every moment since there are natural strange events or created by time travelers that should have change something in the last 50 million years, in fact the expansion creates other conditions every time traveling in the universe in the celestial vaults and the past is not in the same place or accelerate to 100 years ago, it was backwards in the beginning of the universe in the area where it was 100 years ago since expansion takes everything in its way including the time dimensions and all the creation, perhaps something that would be explained in the future by scientists and will be studied by astronomy students, their subsequent physics and astronomy students of tomorrow or the books from the future will clarify it more and science investigations.

The fact is that if we want to travel in the universe all this will have to be consider before accelerate at light speed and get on time to one place, so to speak, since they have to accelerate in other dimensions where time doesn't exist or it stops rather to get to the lunch time in some place or in 10 hours to a week and it would never be practical to travel in the normal universe.

Ok so far so good, but we will have to get into it ever more and it would be like this, the cavities from the future past are "then" or temporal recordings since someone is killed it would die and how can be possible? Well we are here who live in our time, we are in the perfect present and those who come from the future will be here and not in the perfect present which means before something is created otherwise they will keep coming where the recording already exist and is so complex that we would have pages and pages explaining all that could really be, but to travel to the past and the future it had to be that secure and it couldn't be otherwise since things would disappear things every time since there are changes all the time in the past that are anchored in things every then and becomes independent of the past and the other future stays like spinning without future or just in a few and it doesn't create other dimensions at least something happens or perhaps the celestial vault claims it in a natural way by requirement effect and claim of atoms quantity, but we couldn't explain, although, some hypothesis would be found in this chapter, so everything staid recorded and many things would be questioned and will be answered along astrophysics quantum literature only.

That's why we will say that recordings are in thens in the incredible and wonderful stellar NET of the celestial vaults, that's why when someone changes something comes from the past would create another future in other stellar screens not yet define just a little in our episodes and ours will be there forever as the bible says and some theories expose everything is already exposed and it's already written from the beginning to the end perhaps or at least most of it and maybe what other alternate times would be atoms effects and an intelligent universe perhaps more then what we think, but even so we and our work would be there and that's why nothing changes every time and also you have to accept and know we are not the only unique present nor perfect present and that the universe was created 13 000 000 years ago and the first civilizations were born perhaps 6 000 000 years ago and the universe conditions to create life already were there and the stars just were born in this universe 13 thousand million years ago, that's why the perfect present that rules the universe could be something like stellar date of real SUBSEQUENCE ICA1, but perhaps every star has, as we have exposed, its perfect present and times because of its gravitational fields and every galaxy perhaps, that's why we have to start thinking we are trapped in time and space in a small planet under a small star gravity very stable and durable and that universe time is not ours nor the past future but we are part of it and because of that perhaps present time and past and dimensions are there a long time ago, in fact there are theories that the author to relax he ask you if you think future is your perfect present and you are just the past or the past when you were born or we are part of a universal time and that's why we are another link? Of course that's for you to reflection, what is the past future in its present and so you understand dimensional time and also so you don't forget that we are too little to be the time that rules the universe and that's why perhaps many times watch us in this precise time in history itself, everything is historical and present future past.

Many would think that dimensions and the "then" are just time differences between a microsecond and a millionth of second to infinite and our author doesn't think that, for him that's kind of little skeptical and circumstantial in the universe so wide and so complex of so many things that he thinks that doesn't goes with the logic of what really is and also there are existentially samples and prove of contrary and of what is a recording in celestial vaults and also there are evidences of more things that dictate other things like alternate universes and worlds and not just one basic expansion of the universe, perhaps they jump there since in otherwise it would be rare they could be in the yesterday and in the tomorrow at the same time and just occupy time space, they wouldn't fit in many things and they are expose in this book like propulsion and dimensional jumps of space travels, if going there they couldn't just be small times or is it? The author doesn't think so, however we could leave it for reflexion for those skeptical and perhaps also dimensions too, although the author doesn't totally deny it and just deny it and say everything could be written in other more complex in an existential recording in a hipper space of a very complex multi universe.

CHAPTER 4

Real Time continuation V4

EF-17

Back to the astronomy studies from 8 decades of the best astronomers it was discovered and is indicated that we live in a universe in expansion with millions light years of distance from one star to another and somehow that universe was formed with a big explosion approximately 13 billion years ago and because of that, galaxies are traveling and expanding and at the same time they take all the suns inside of it rotating on their axes and also they go in such expansion to one outside, that at the same time is incredibly big, so big that with our modern instruments of the 21 century nor radio telescopes in orbit there is no way to see the beginning nor the end of the universe, that means is not possible to see where it comes from and we are not able to know how far that expansion reached according to recent reports and calculation maps have been drown up of it, but real images of modern radiotelescopia, is the best way to see even through gases and nebulae, is not possible to see the beginning nor the end and in those 13 thousand million years that scientists calculated the universe had at that time all the elements couldn't have formed and just in the beginning where the big bang started that explosion ejected all that material there they could have formed, but we have found other explanations like the possibilities of several big bangs existence and so several universes and also in this book we will expose there are and must be several universes that perhaps started in other times and this one in other or when exploded they were formed.

On the other hand, we know the elements table that form matter are 119 and sometimes one or two new ones are discussed but that's it, so if the suns can create 3 or 4 elements according to astrophysics studies since there have being time only for 20 to 30 elements since it takes time to stars to create elements, as we know ours has created like 3 or 4 elements in the 10 thousand million years of existence and also the other suns with less time, so if we get together all the 13 000 million years it only would enough for 15 to 25 percent and if stars were too fertilized then from where the other elements comes from is from other universes or from a mass of matter there was and exploded but following the routines of matter recycling the universe has, we believe it comes from previous universes from the big bang and have being there for thousands of millions of years and only studies of the future or from science of tomorrow will guarantee it.

So that's why we see there has been not too much time for that many elements and we believe they come from this same time but from a previous universe to ours that expired and exploded again and even more before this one and maybe they really come from several universes like from 7 or 15 universes since this one lacks the triple of life until fuel hydrogen and helium ends and stars turn off without fuel or perhaps until they get lost in infinite creating a celestial vault that maybe one day it will stop and that would be the end where there are no atoms and precisely there must be most of the answers, since there we could see how many celestial vault have been created and see the how many edges, as long as we have a spaceship to go there and see how many vault we can see and if matter someday stops and if the first explosion is still going on still the question is and how far has created a universe, how far there are universes and even more how many parallel universes really are there and dimensions, since the answer is we live in an incredibly big and titanic multi universe never dreamed, so huge that we could go a life time in the best spaceships ever and even if we reach the end we must see all the ends to be satisfied but maybe we won't see it and we could see dimensions in normal dimensional angles several layers of several universes in that end of what matter pushes in forms of galaxies to a places where it has to be an end and we could see an edge of a book with several layers where we'll introduce some areas and several older diverse universes, of course the author deduces and exposes in this volume that his studies about this lead him to think it's not just one edge and there are many of them and there we could see what end has, after all the matter that has being expulsed to an improbable giant end that seems endless and where perhaps we must go further from the first galaxies and travel thousands of millions light years to places without matter in an incredible darkness and where the expansive wave goes making way to places where if there wasn't a universe before or ended first there would be only the Nothing with no atoms and the real last edge of the universe of our big bang and on the contrary if there was a previous universe we'll have to go to the end of that one and the other one traveling thousands of millions light years to an strict end where maybe even time would stop somehow since if there is no expansion everything would vary in other ways and maybe there would be other exotic physics laws and other laws would rule for sure and even so someday technology would take men from the earth over there and it would be marvel and amazed of seeing what is the end and the beginning of the great big bang, or all the previous and posteriors universes to this one full of surprises like physics and technical that they would have by going there and even so some beings born before and others over there and in that end they will know what is there and traveling by through those confines they could go into several layers has we previously exposed and several dimensions and only counting them in those future eras will be known or when our spacecraft go there.

And so returning, to our theme of time

EF18

SPACE EMPTYNESS AND UNVERSE GRAVITY, HELIUM AND HYDROGEN EXPANDED AND ALMOST DISINTEGRATED.

By this media we expose hydrogen and helium are what most abounds in the universe, is the space emptiness or components that abound there but very expanded, as if its core was at one mille, and even is not like that but resembling the air that can go through the emptiness and it has atoms of it since the expansive wave of the explosion that expanded the universe "the big bang" create it and also many other things for sure with the push wave of the universe expansion and our author thinks GRAVITY IS A POSITIVE POLARITY OF SUBATOMIC PARTICLES THAT ATTRACT AND CAN BE FELT for example is like a grenade explosion that throws things with gravity waves, so you can have an idea more or less what is what haves us standing on earth and it's known here on earth they are positive negative particles at a subatomic level already discovered and dominated but yet to isolate and catalog, and in theory its known only that could be but there are others already isolated that science already cataloged but not all are cataloged yet, we are just starting to see that micro universe or subatomic particles tapestry experimentally although they affect us daily significantly but science lacks progress in it and we dream and want the gravity particle the famous graviton and perhaps it doesn't exist as we believe but as a series of magnetic poles positive negative of attraction of several subatomic particles between the normal particles of the atom that has the strength of gravity as acting together, that's why the author exposes the expansive waves of the universe that's all they are since they cannot be anything else and also he expose the space emptiness must be expanded atoms of helium and hydrogen and other elements, so in this book he explains the electrons are separated from the atom like at one kilometer or meters away as we mentioned previously and there are other gasses in the atmosphere even tenths and we walk among them like oxygen and nitrogen, so that way the space emptiness is just an expansive wave of helium and hydrogen most of it, highly expanded at its size and perhaps they are other type of atoms like if it was an expanded spring and gradually stretched and that's why it looks like space emptiness but actually there are many things there and at the beginning formed dimensions and parallel universes of different density and different times maybe, since the expansive wave as it exploded it wasn't uniform and created several subspace layers and different times.

Therefore we explain to you the following, although they act differently in the atmosphere than in the space emptiness, that's why we dare to say there was a universe before this one and there was already a celestial vault, the big bang acted as an expanded wave over something already created, which means, over previous celestial vaults otherwise it would be different, anyway there would be that expansive wave of atoms that throw matter all over the universe and created expansion but the most peculiar thing is it created the space emptiness and today only a few physics have explained it in detail really but very little about what it is since everyone are limited to talk about other things, but studding it well what it is, we define that is just that, expanded gases and other matters where they can be seen like emptiness otherwise there wouldn't be gravity and planets would be adrift and they are trapped in the magnetic swirls created by those expanded elements perhaps helium and hydrogen in their subatomic particles and those swirls have earth trapped around the sun and the sun around the galaxy and the galaxies in expansion in an enormous push and that positive negative magnetism or gravity is created from those elements hipper expanded traveling in the expansion and also of several layers time space to discover. And because of that continuing deepening in our theme, there are others that speculate they could be other gases, but in fact that can only be expanded gases to the extreme like a trail the explosion and expansion left as that emptiness exploded in the first universe when only existed the NOTHING or in the second universes and third and so on, regardless of how many universes have been before this universe and there are speculations of several or a few but regardless the author defends that the space emptiness are several gases and some expanded elements and their particles that behave in several ways and create the celestial vault and perhaps the atoms are in the orbits separated cores of the electrons, to understand, every mile and its electrons perhaps changed and they are strange, maybe other parts like stellar particles or just parts of positive negative subatomic polarity that are attracted and that is the easier definition of what is the space emptiness and the celestial vault, in a few words our home where we live, everything is to be scientifically proved by tomorrows scientists and their experiments GRAVITY. Even so, it has to be cataloged what more can be and how many more expanded elements are there, their miles or meters might be more elements or the same as the stars in different proportions, science will go dictating what elements are and that way could clarify little by little those enigmas, we expose it like this to explain time and space in different times and same spaces, but returning to space emptiness since those are very expanded atoms defined as emptiness, that is, we can go through them in an emptiness but is not really an emptiness and here on earth happens the same thing with gases has oxygen and nitrogen that are expanded in a much minor way and we go through them every day just by walking. So because of that we will see that gravity is something that affects polarity between the expanded atoms from the space emptiness in its positive negative and that way we could study easier about gravity and its behavior and the gravitational forces of the stars, the planets and the galaxies and their rotation like the spirals they create and trap the planets like the one of the sun to the earth as they rotate and the sun in the galaxy and the earth to the moon, that is why once again we define gravity is the force of polarity positive negative of subatomic particles not discovered yet and others attracting to each other and also in the swirls that create the stars by spinning and they trap celestial bodies has planets and everything is between the atoms already different very expanded because of the universe expansion, hydrogen and helium atoms most of.....

----------------------------------00----------------------------------

And so in universal time

V ETA CARINAE A1 Y IC A1

In this occasion we will explain more about the cataloging of alternate dimensions and alternate universes and also as we exposed before in detail for your better understanding of the orange slices theory and it would be eta carinae A1 which means that in straight line to Eta Carinae star with reference A1 this dimension, A2 next dimension and A3 next dimension and so on until 64.

Even so, changing universe and not just sub dimensions of the parallel universes, we will use the denominations of IC A1 IC A2 and so on, but we will do it to the extent of importance of the dialogue like if we use Fahrenheit degrees or centigrade only since sometimes we will want to say "in the next dimension" and "in the next sub dimension" eta carinae A10 and sometimes we'll need to say IC A10 and it would mean the same thing, but the only thing is that the denomination IC is has more importance since is a reference of more distance since it is at 1000 million light years from earth and Eta Carinae is at 7500 light years only, so we will talk like that because those denominations came out while designing them since it was needed something farther as reference and sometimes something closer.

All because we cannot see the beginning of the universe with our telescopes from the 21st century and neither can the end of the universe be seen and as we have explained not even the pull of return that the universe has in the beginning we can see, that's why we use a practical galactic measure and a practical universal measure of a thousand million years in straight line as denomination ICA1 which means straight ahead to the galaxy IC1101 and which is the biggest one discovered with 6 million light years of size like 60 milky ways and it's at an astronomic distance of a thousand million light years, since what we are explaining is to delineate alternate dimensions of each universe and its limits.

Also sometimes we will use vortex A1 or vortex A2 to delineate the limits of the universes not just from dimensions, also we will use the words and definitions Big Bang A1 and Big Bang A2 that sometimes appear as a reference of other explanations only since with those we are going to work most is eta carinae A1, 2, 3, 4, 5, etc. and ICA1, 2, 3, 4, 5...

Also there would be other explanations and denominations for future times and pasts and would be added to eta carinae FA1 which means future in a time in dimension A1 that means this dimension of our perfect present and so we will mention it for more important dialogues of the same conversation ICFA1 and ICFA2 etc.

and for the past we will mention ICPA1 and ICPA2 like eta carinae PA1 and so on.

We also are going to expose that there would always be alternate denominations for other things like several time space that appear in the studies or conversations of alternate future and alternate artificial dimensions so we can travel through the universe or rare dimensional places like ICB2 or ICH3 and ICFB3 or eta carinae B2 or eta carinae H5 since every dimension would have variants or each universe more.

For the universes we can use ICA1 and onwards and Vortex A1 or Big Bang A1 tec, but not eta carinae A1 since eta carinae would always be for definitions for dimensional conversations less important or more personalized, although it will be at the discretion of how you define it and other measures can be used like UT A1 and UTA2 that means universal time A1 and universal time A2 UT A3 etc.

And is that we will be defining extra dimensions and extra universes and they would give that denomination if they want to, but it would be more useful this other denominations to study time since they define more exact things and about the theme we are talking about of where is located time and where are alternate dimensions that occupy same space but in a different time and the distances in which they are defined and how you cannot delimit with the word Time those things just like that and also we can't see the beginning or the end of the universe with a fully identified reference and we can't either see the universal time to delimit the exact bordering when starting to study time because is easier for students to learn.

And as we said before, they would be corroborated by tomorrow astronomers and physics.

And so in universal time

EF19

them and the expanded atoms that must have something like in the atmosphere there are other gases not discovered of gravity and as creating differentials of polarity positive negative of subatomic particles create gravity magnetism and trap the rest of the matter and in the universe in the stars like some swirls that have trapped planets, suns and stars in their gravitational fields.

So that's how we define the expansion itself and also its orbits in relation to the object that influence, like the earth spins trapped in a gravity swirl that the sun creates and the sun in the one of the galaxy and the moon in the swirl of the earth, are GRAVITY, since as they spin influence those particles mentioned before in an inertia and attraction positive negative of multiple atomic stellar forces perhaps or from the existent elements in the planets in their subatomic levels and in the universe its expansion, everything that affects and lives in the great Celestial Vault created by expansion itself.

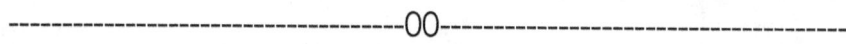

PERFECT TIMES

Perfect present, perfect past, perfect future and others real.

Existential times, Parallel Universes and Parallel dimensions.

By this media, we dare to say TIME is like different strange and exotic things in a net of the creation that are only glimpses in modern science of our civilization and because the fire era was 200 000 years ago marked the beginning of other eras and perspectives and tools for the homo sapiens and the Neanderthals, for us time and gravity will be that fire that will take us to the stars and many places and will create us endless benefits, which means, it could take us 2000 years to dominate everything without help from the exterior perhaps only centuries, since 150 years ago we were on wagons pulled by horses and today we go to mars and record videos with our androids, flying at supersonic speeds, also at that speed we go around the world every 90 minutes in the space as the international station does and maybe we will go to the hardest time in perhaps 150 or 300 years more, we don´t know, but to dominate everything it would take perhaps 1000 to 2000 years, on the other hand maybe we will communicate with dimensional frequency transmitter, which means, when we make vibrate the space frequencies and make contact for the first time and stimulate molecules from parallel universes from where our signals could evade Time Space since we could tune on some radio that could give us formulas for short cuts from other civilizations or perhaps if our signal get to a place where there are beings that could give us a jump in technology and take us to the stars sooner than other things, that´s why we should study time since is what rules the universe, also is very exotic, so in this occasion you will have the opportunity to observe why we don´t say the author is an ally of the existence in many ways of the past, the present and the future existing at the same time in a more complex universe creation than we thought, also it exists in several different shape dimensions that´s why many worlds where some were in other shapes or were changed by chrononauts and its part of this book chapters we already explained and JR rules dictate and those are nothing can be changed if you travel to the past since everything exists and eternity of microseconds to infinite since nothing has been studied really and nobody knows if that's how it really is or we would see frequently people disappearing or burned woods appearing because we would be very fragile if we were only the remaining of thousands of events that happened in 50 million years repeating it and previous chapters say it if in 50 million years nothing changed or no chrononaut came or no time traveler it would be too difficult that nothing happen and that time itself wouldn't change in a universe so complex of chaos in expansion and it travels at a great speed and where very strange things happen, which means if it can't be changed and if aliens or gods take care of us it would be also very fragile to depend on them and the universe would be superfluous and weak and it is the opposite is strong and titanic and in a nutshell it would be very difficult everything be that way, already explained in chapters of this book and if everything could change us and erase us but evidence tells us is not like that since nothing ever changes and nobody disappears before our eyes ever nor buildings or forests, there they are and nothing could change them in the past, so they won't disappear in front of our eyes, but that should be read in other chapters of this book, since is only a reference to explain the present and the future here and because of that we dare to say the strict present is the one who rules in importance in things and people and personal strict present is when we are born or when we are conceived and perhaps when our grandparents had relations or when a previous generation was guaranteed or just when we are born since they are variant that we will detail later but perfect strict present it would be that one and of course the one its directly related with DIMENISON IC A1 explained in other chapters too and those who follow and the other perfect present would be the one we are living in the time we are and perhaps there are time laws not studied yet that say strict present in this case is this 1CA1 or in the minor denomination Eta Carinae A1 both invented by our author and exposed in other chapters.

On the other hand, perfect present for the universe is the one who rules the importance of civilization but in relation of a person we will say and simplify it begins at birth and the other one where we are, although there are other presents like perfect continuity of the person in the dimensions eta carinae A1 IC A1 STRICT without variants and then the other presents that are in the first dimension of this time, also we are in the dimension that rules exact civilization with respect to the universe and existential expansion and the one nothing can change since it was created as a record until the end in a hard disc of the universe, which means if you want for the skeptics the Then of or the record, as the astronomers see physics from then or of universal record at least someone had changed it and has the technology to change perfect present and arrives before you get there being in the past and didn't created all the Then or perhaps it did but is not convenient to no one to change those things in that technological level since it will affect a lot in the future and someone would complain but it would be necessary a special ship with a special weapon to get there before the perfect present and annul that person and erase all of his legacy in thousands of microseconds to infinite but such thing is impossible since there would have been created 80 years of that person and would have to erase for a reference every microsecond to an specific person and it wouldn't be possible because being born would exist in time, the 80 years that would live every microsecond to infinite or perhaps every million second and there would be in 80 years 60 000 every minute and it would be there before you and there also will be the all others, so all that stays written when perfect present of things begin millions of years to the future and the its consequences and it only could create another alternate future explained in other chapters and the record and the then or several then with the person there would already exist there, but such thing is controversial since it could be a multiverse if it is a record and if not the other way than continuous and that way our atoms recorded in the universe atoms or created then, but observation tests dictate nothing has changed in 50 million years since the same forests are there without changes out of place like disappearing in front of you and the protagonists of history are the same and its consequences and no person disappears in front of you, but other will argue is what was left of our history and our author argues is not possible that in 50 million years 50 000 years no one had come through time and change nothing specially when is not the same universe every minute and because of the expansion we are in other area of the celestial vault and those could be natural effects of other kind changes like gravitational rays and civilizations war that someone changed or thousands of chrononauts, that's why we know is recorded and nobody can change it, only working years on it, so if someone gets killed in a few seconds it would be right back again if you go to the past or maybe just cannot be changed and only the psychics that are with us in the perfect present can change the future but perhaps they are directly connected with us in this perfect present or rather they do it looking at the future from the perfect present which is here and it didn't came from tomorrow.

On the other hand, the perfect past is when we are born and the one that already happened and perfect future is the one of tomorrow that hasn't been but is already there since if we are HERE perfect future would be there too and no one can change it at unless he travels to the perfect past and avoid something and come back before the electromagnetic record subsequence is created of the great celestial vault or failing that THE

THEN OF PERFECT FUTURES every microsecond to infinite and perhaps every nanosecond to infinite, which means, every million second, but simplify things, we will evade time and we'll say is every microsecond to infinite. That's why, if no one changes and no building changes and doesn't disappear or if a forest doesn't disappear before our eyes, is because nobody could do it in 50 million years and nothing changed and that is IRREFUTABLE AND IRREMEDIABLE and because it is a very complex universe, impossible to galaxy full of wars maybe and of UFOs traveling around there or perhaps deviated cosmic stars by changes in the expansion and temporal effects of nature itself, but accepting time travel exists and that we want to go to galaxies and stars and that we wanted to change history, we believe always warn that then where a person exists unless he has the criminal technology to change all those thousands of Carlos and of Elisa it cannot be done, if someone changes just like that in that present, its known perhaps exists what we expose in the chapter about it in this book and it would be cloned and would be other alternate time and maybe it won't get very far but the first time ICA1 it would be there existing into the future.

For those who don't understand we recommend to read the firsts chapters and review it until you understand; so anyway, for perfect present could be at birth or being there by consequence of the eta carinae A1 strict and subsequence of the events and it only would be the subsequence of what already happened, and that is what is next to remember you to understand everything else.

The creation of one thing or a person creates futures every microsecond at a speed that would be defined later, but it creates them and there they are and that's why we know nothing disappears ever and is because in the past even if something changes the then as they create or record in the magnetic stellar cloth, explained in previous chapters, from the past and the future it creates them and we are practically an electromagnetic record or for the skeptical, we are a creation of then from yesterday today to the future since we are electrons and the universe is electrons and we could be taped in that cloth has we expose it and it wouldn't be possible to travel through time and there would only exist a trash of a time to the future and temporal travels wouldn't exist ever and nothing that we see could be, neither go to the past or future and that we'll only know it if we go there, has we know by experiments that do exists and going back to the theme, it would be strict present ICA1, 2, 3, 4, 5, 6 and so on until the future scientists catalog them all, and only a temporal record and affect someone changing the past of perfect present from that then and the one who could get there before the one was there with special machines of time travel that would be perhaps very hard to get and would be temporal criminals maybe, also they'll should go very fast and get there before you since everything would be there before it gets there and only if it gets there before the then appear they could change things that are not possible easily and because of that nothing changes and everything stays the same and only if we belong to the perfect present, which is this time and be from here or failing that, being related it could be for very important things, perhaps everything is protected by natural laws or superior beings or gods or something like that, but traveling normally to the past everything would be secured and traveling into the future they'll always be there too, every microsecond to infinite and when someone achieves to change something be because it would be related to technology that can change it or perhaps is part of strict present or failing that there are possibilities what has being changed doesn't last and perhaps stays in a stopped time and crashes and stays like stopped, and it was like bases of ICA1 or eta carinae since maybe it doesn't has the bases of being a real solid future unless all the changes have a lot of stories.

Thinking deeply you will find out or will be stimulated to study it you time students that read me in my future I salute you and from here in the past I confess you will be getting into time right know to be delightful with it and come with the same conclusions when you study it, many times even better and you will make laws would be stimulated to it since our civilization would depend on the study of it since time consider world population and the attacks of invader beings could do and the way we would have to be protected against them in different time space and even to protect ourselves from natures blows and if we want to go to space they'll should not study how to change the future but to study it cannot be changed and it has to be respected, but we would be able to travel to the stars evading time and space, that's why all those rules of time and those exposed in previous chapters will help you and would be the laws of tomorrows time that will dictate your laws from the rules of JRC the author and I hope you'll like them and satisfy your doubts of time and space.

On the other hand, returning to the perfect present study, these rules dictate perfect present can affect the person and is from base dimension ICA1 and its subsequences and the others will be the next ones but that one will rule and the perfect futures would be those who reach the end of times and the perfect pasts would be those that are at the most possible past and let's say one million years and to the future also one million years and it would be FM A and PM A as the temporal graphic show it.

--------------------------------00--------------------------------------

And so returning, to our theme of time And r to the theme of dimentional astronomy

EF20

Dimensional antimatter and steps to other universes and dimensions.

To detail dimensional matter and who the positronic edge is the limit and specially particles and the new elements there are and will appear in their subatomic positronic particles skirting other dimensions and where matter only appears when our molecules change atomically automatically and where they would meet at the area of positron 1 positron 2 and positronic area 3 or subsequent until finding the required universe by going into those atomic areas and with different values when those dimensions and universes find them and perhaps would be in level positronic 10 all the way to positronic 20 that is 10 times antielectron antimatter and there would be the limit of the matter or just when subatomic polarity of gravity is found we will reach the limits that rule this universe and there other universes will rule or maybe we just have to throw with gravity forces all the particles of this dimension aside in all senses like using and inducing those particles of let's say magnetic positronic gravity when using a metallic sphere everywhere loading it with other protonic values perhaps and perhaps because of that the UFOs are round and maybe that's what the aliens do.

And activate and prove until we get to the desire dimension, in theory that would be one of the ways to get there and to do that we'll have to practically eliminate the atoms with something that neutralize them like neutrons or several charges the way the positron appear in magnetic fields but this time when the we disintegrate the atom as we wish with science weapons of the future and it would be when approaching to the walls of the edge of matter and because of that we have to observe very good the electrons and discover what is between them really but if nothing extra is there from what science have observed then we will have to create close up conditions to the positrons a, b, c, d, e, f, g, that haven't been discovered and those are the particles that appear after the positron in its same structures and only in laboratories of modern era appear but they do appear exposing the atom of several materials to magnetic fields the positrons manifests that really are the antielectrons, but at the beginning of the experiment if normal electrons are placed in a container where exist magnetic fields and change the environment in a magnetism of magnitude explained in the experiments of particles accelerators in a special chamber, there they have photographed them and value them and observed in what they transformed in and they transformed in antielectrons or better as matter catalog them in Positrons and therefore we believe that other variants should start to emerge from there, although we know is a polarity change it does not stop being strange and is the road perhaps to visualize and start to see the way to alternate dimensions and the vortex of parallel universes and that maybe we could delve in observe other elements like the gas particles expanded in the space with no atmosphere, perhaps is easier since they are very expanded gases explained in this book and there we could see how they react as they change of environment this kind of material from the space emptiness and when we perfect it perhaps it would be necessary to keep going with the materials of the most compressed elements like the elements of a planet like the earth, perhaps they are easier to physically manipulate that's why we should go to the magnetic random calculations from each observation and that would take the scientists to the next steps, because doing it without some instruments that diffract the atom without particles accelerators it would be always difficult to find the others and we will just deduce them mathematically perhaps and an example could be creating something that dismantle easily the atom and that would be with induced neutrons because it would neutralize its charges and other would emerge or would liquefy them neutralizing its charges until see what happens, but only scientists could know how viable is to do it today because as the charges vary and neutralize the particle with some neutron gun or of what they have discovered already, the antineutrons perhaps that is the way to get to the other particles. So as we see those horizons we will get into universes of different charges but the sphere would be needed, our author propose that and it is a metal sphere with gravity induction in all senses until it disappears from this universe just because rejecting all the atoms around it and that way appear in other universe and those experiments can´t be done because first they have to induce gravity particles and first create gravity or antigravity and studying that antigravity they could perhaps focus it to get out of this universe but since our deductions take us to say gravity is polarity of subatomic particles from space emptiness and in planets of the atmosphere compression and terrestrial attraction from atmospheres and we have to go deeper into those things and the first step it would always be the discovery of all the polarities of subatomic particles that are involved in creating gravitational attraction so that way reject it and push the other dimensional systems or failing that, the search of the most wanted particle denominated and not yet discovered the Graviton that perhaps has we know it doesn´t exist and they just are a lot of subatomic polarities from several particles perhaps positrons a, b, c, d, e, f, g, etc. and antiprotons antineutrons 1 2 3 4 5 6 etc. and their interactivities. The end...

-----------------------------------00-------------------------------------

CHAPTER 5

And so our author exposes in the
subject of temporal astronomy

EF21

H2O SPACE WATER

Much has been said about H2O or just water, something
life is sustained with in this planet, but not necessarily is
sustain other planets and environments, here on earth
there are many microorganisms that doesn't come from
oxygen and is not needed for life exclusively because
other being use other ways but when perhaps they
become mammalian or something like us, in our time,
that is at a medium age going to the second or third part
of a planet, perhaps they do have to be from oxygen, but
water isn't just oxygen but hydrogen and oxygen, and
that is more than that, water is a strange liquid, abundant
and transparent and why is it here? Many scientists ask
and what brought her here to earth in those proportions

unlike the other planets, well scientists explanations have been until now let's say slight lack of science and sometimes even childish, like meteors brought it here and the author of this book doesn't believe that and denies is that way.

Because on the other side it would be too easy and he believes the modern astronomers that deduced incredible things such as documenting how it is the dead of a star and what elements are formed like iron before collapsing and they marveled us with that and is incredible they have those explanations of the water on earth that is like in an orange, the skin of its total mass but abound in its surface a lot and he thinks they just didn't had time to see other theories visualizing the cosmos light and the radio telescopies and suggest because of that water was created in the explosion of the super nova mother of the previous sun to the actual, in its fusion at the end in a nuclear fusion explosion and have traveled light years of distance and then it concentrated only on earth, is a weak theory but believe me is the only one there is and is not the best from the other explanations, by the way very poor ones that explain from where came so much water and not on other planets.

That's why our author exposes he has never agreed with it and he has always thought that perhaps it was a kind of solar elements or underground fusion that were liquefied with time and oxygen and hydrogen were able to be together and merge in a NATURAL COLD FUSION OF THE PLANET here on earth and other planets since they didn't have depth pressures, something like that, that have petrous products that have water trapped in its molecular structure like volcanic lava perlite and it's used in agriculture and gardening and already expanded in an industrial thermic process that free water but on the other hand it would have to be a lot of examples of it and there aren't, but even so, we know it could be other things like minerals already transmuted in cold of primitive earth.

On the other hand, the author elaborated a theory from himself that dictates that the sun hydrogen was thrown by flares being unstable in the beginning and that's how it was expelled to space and in the explosions of those titanic flares IT WAS MERGED WITH OXYGEN FROM THE SUN AS THEY EXPELLED TOGETHER THOSE ELEMENTS when the sun expanded in very strong strange behaviors and sometime throw hipper giant flares that could reach earth easily and that's why we say the sun send to earth an electro plasma at huge temperatures of millions of degrees Celsius in the right amounts to create fusion of oxygen and hydrogen so that way as they precipitate to a gaseous giant atmosphere with different pressures created water and that is why the layer that was fused precipitated to the surface and created the seas of earth and united to the rotation of the earth that was perhaps to big compressed it and the rest of the layers of the huge atmosphere evaporated and went to outer space, already back then earth was a gaseous planet like Saturn sort off.

On the other hand, that fusion could have been created in cold or heated by that flare that lasted a period of time since the sun has all the elements and maybe lasted one day and was trapped all that water here as the earth rotate and trapped it like a roll paper in big amounts here on earth because of the closeness to the sun, also it could have been, as we said, other things like meteors that crashed with the earth at great speeds that created or triggered a thermic fusion and the element was made that leaved with time or was turned into water. The best possibility united to the flame of the sun, is this one too that we expose next, that's why he agrees more with one theory he thinks it could be the one and is that the water came because once the planet earth was once a gaseous planet enormous giant something like Saturn and since it didn't reach the mass of Saturn it made it shrink in its own gravity and all the gas atmosphere like hydrogen and oxygen and nitrogen were in strange concentrations where the faster rotation perhaps and gravity were great to CONCENTRATE AND COMPRESS such atmosphere of hydrogen and oxygen and that way was merged in cold or having atmospheric pressures of a gigantic atmosphere never imagined of the earth in modern science in the solar system, in fact it´s even very possible because if there are more gaseous planets then solid rock, in fact there are only 4 rocky planets and the others gaseous and those are Mercury, Venus, earth and Mars from the nine planets and also small and not significant like Saturn, Uranus, Jupiter and Neptune that are giant and very of great significance really.

Compared with earth and he thinks earth could have been a gaseous planet and the gas of our primitive atmosphere was precipitated and concentrated and was compressed and merged in cold because of pressures and the faster rotation of the earth and closer to the sun and at the same time be more abundant in gas and hydrogen with oxygen that obtained from the primary cloud in its formation and when some meteors crashed the explosion merged all perhaps or sun flares of liquid plasma or simply was merged as gravity compressed it and that way seas were created and the compressed atmosphere staid there converted in salty water; or pressure having more gravity as spinning a lot of times faster in the core or in the rocky surface of our planet by a series of rotation and gravity processes of the gaseous giant atmosphere itself but with a deficit and minor to the others from other things, it concentrated and made it precipitate in the surface of what it is today and with those pressures and with the rotation movements and gravitational attraction that maybe they were tremendous and create thermic processes of friction of the core affecting the layers of the gaseous atmosphere and managed to precipitate and fusion or get together hydrogen with gaseous oxygen and that way create abundant water in the planet and precipitated creating the wonderful oceans and lakes of planet earth... but such thing scientist would have to prove it someday

-----------------------------------OO-------------------------------------

And so our author exposes in the subject of temporal astronomy

EF22

BLACK STARS, COMPRESSION
HOLES OF ENDLESS MATTER.

Much has been said about the famous black holes and has been marveled, it has also been said about galaxies where this phenomenon abound even today and it has been discovered and proven with radio telescopic has you know a super massive holes in the center of galaxies created by hundreds or thousands of stars that fell into gravity centers of other old black holes and got bigger joining other hundreds or thousands of holes and stars swallowed by themselves and it's known they created the hipper-massive in the centers of the galaxies.

It's known there are pulsating stars that created other phenomenon that didn't turned into black holes and staid as white dwarfs or pulsars phenomenon that as a ballerina spins in its own axis, sending some intermittent beams of light that never stops like a non-stopping lighthouse and are those that orbit their own orbit a star that is been slowly eaten by the other one that transformed in a black hole or failing that in other cases of binary stars they stay spinning to one and other and in this case are the most frequent pulsars and it's been said about this binary stars swallowed entirely by black holes because of been born together in the universe, on the other hand as we say there are white dwarfs stars that are stars that were left from suns not so big like our sun and magentas and other ends of stars that are left after their nuclear combustible of millions of cubic miles of hydrogen incandescent in internal nuclear fusion ends. We also know there are Quasars and they are what black holes spit in form of energy streams and they did that when they reach their mature age and after devouring a lot of matter and couldn't resist more and our author ads and exposes that a cause according to his studies and logic deductions of the celestial vault explains us that the age and time of explosion of a Quasar is determinate by the way was deformed or changed with the explosion in outer space and its reverse wave that created the black hole, something that he hasn't explained before but in his thoughts determinates it is because all the explosions have to be different since all the galaxies or giant black holes have different mass and because he doesn't believe a reverse wave or rebound to the explosion of a super nova could be endless but perhaps could be compression wave through the center and perhaps other astronomers agree or not, and the compression waves are more logically so powerful to create a black hole and not a weak pull wave or counter an explosion only, also he exposes today he created the dimensional theories from IC A1 to ICA64 to 100 since he was convinced there are several dimensions and other parallel universes since is easier to believe that is how it is a QUASAR since every explosion would be different according to its amount of matter of every star and specially the Quasars of the center of galaxies because they must be not because of a return wave but because of several parallel waves from several series of walls of the universe in this case perhaps in the center of a galaxy as hundreds of stars and as series of holes merging since there must be other forces although from the universe walls and perhaps from there are the spinning waves of the gravity spirals of the galaxy that imprison like a twister the hipper massive holes that are really different but they are black suns jointed and hundreds of swallowed stars, but in this case is not just made of the return waves but also the spinning gravity waves from all the galaxy and the laws of a black hole are more strict there and from which little has been said due to they're not visible from here only with infrared since they are too far away and covered by thousands of millions of stars in the centers of the galaxies but sometimes we see the famous power jets when they collapse in a way that they throw those jets in their opposite poles in their opposite poles from the center of the galaxies and studying the orbits some scientists came to the conclusion that they must be hipper massive black holes that makes them fall in gyratory orbits through the center of the galaxies so many stars and matter and that's why they spin like that to a center of galaxies studied like milky way itself, in fact one day in the future a power jet will explode called gamma rays explosion and that is what really forms and advise from a QUASAR, although it has not been determined when is going to happen and is not very sure if all the galaxies have that phenomenon some time in their useful lives, but is something not very studied since happens every hundreds of millions of years but even so perhaps wouldn't affect the earth since we are in the edge of the galaxy and that affects the centers of the galaxies only and what is in their axes of reach.

But what they really are? And what is a black hole? Nobody can establish it with exactness what are they and is because modern science is focus on observations of lights and electromagnetic waves, although they have marveled us with explanations that change while they advance, they thought they were better like exposing life of the suns and other ways of astronomy of latest generation, but knowing exactly what it is only when our pathfinder get there or when most of the scientists really agree and closest is what modern physics explain and our author tells us other ways and new angles for you that perhaps you like with the parallel dimension theories and how they are formed and try to classify them for you in a preliminary way.

On the other hand is has been delayed or been neglected at least in their writing and explanations the logic deductions of what they could be, also they have not been able to gather enough information some astrophysics writers or willing to talk about it in books since they think it has to be scientifically proven but not what is so far away since it couldn't be that way perhaps because of their strict careers and because of that they have forgotten about the angles to reflection about the things they should explain more, like those the author will expose along this book so the astrophysics students of tomorrow can be stimulated and won't have limitations or at least of thinking ideas or scientifically untested theories or even they could test them just by thinking in some things and have time to think in others and that way save some time to modern astronomy and perhaps our author had the opportunity to reflection more about those options of what is up there in deep space.

Black holes are next.

And that's why we expose that the black suns are endless compression of matter created by walls of a extrange Stretched and contracted SPRINGS big coils, of subspace layers, deformed by the super nova explosions and we do not agree they are just reverse waves, By the way is a proof of the existence of other dimensions of side universe, and in the hipper passive black holes in the center of the galaxies we think they are compressions of gravity rotating waves of the galaxies eader, and gravity spirals and of course thousands of suns and stars falling in their gravity centers as other scientists have exposed and in the case of black holes we think they are created the giant explosion of some stars as we all know with enough mass they can expand matter in the universe, and in this case we dare to say what is it and what are the deformations in the walls of our celestial vault in alternate dimensions and parallel universes that affect such explosions getting to a limit of expansion of the matter in a point of percentage of the expelled by the explosion from the parts of the star ejected to the universe once and how almost all its matter and gases of what they are made of and it's even known there are fusion and thermonuclear reactions where many scientists think elements merged like gold in the lasts minutes and that way they are thrown to the universe in the last sights with all the hydrogen matter and other elements and incandescent gases from the millions of cubic miles of what it was as a star, between those elements the most abundant hydrogen and helium which would lost balance between internal fusion of the exploding star and the expansion of the fusion in their explosion and as they throw that material with a few bursts like the nuclear throw large amounts of materials and consequently throws GRAVITY WAVES outer space.

And by being able to eject quantities of matter more than normal stars since it throws also gravity to universe more than normal since every body in the universe creates gravity and when a star explodes is a titanic push of gravity never seen and perhaps only surpassed by other forms of black holes, bigger once, like those from the galaxies in their center, the massive holes and which we haven't seen like those from the beginning of the universe or the end of black galaxies or those from the great attractor area if it is a super massive.

Well, going back to the star that explodes with enough mass to create a black hole in a supernova explosion, as it does that it throws, as we have said it before, millions of cubic miles of material to space and in a moment modern science says in that explosion a part comes back and most of it moves away forever from that place at amazing speed and the other part FALLS IN DIFFERENT EFFECTS that we will explain next, since it's known it obtains a gravitational wave of reverse that returns expelled matter back inside and that creates the black hole.

Certainly we believe that in the beginning of the universe something like this happened, creating the parallel universes of this universe, And of the former ancient universes, to this. And its diversities.,

The rebound wave creates that great anomaly as they call it in the texture of the universe called black hole or black star and so in this volume we have explained several things that exposes what happens with them but we will synthesize some, and the author believes based on his theories of dimensional time and alternate dimensions and that perhaps the walls of the universe get damaged as the supernova explodes and is affected with the expansive wave as with the reverse wave and the universe ends unbalanced in its hegemony with the other universe that's why matter penetrates with such power in alternate dimensions perhaps, that's why gets through walls and dimensional scales that affect and alter our universe itself and also they connect in layers and of layers of walls of the universe or alternate dimensions between them, that's why they create a kind of reverse spring incredibly titanic, forever poorly accommodated because of the supernova explosion that creates a rebound effect of endless dimensional connections through the inside of the center of the black star and which doesn't have an end because of the spring itself already defective since something like the explosion keeps stimulating it and making it act like that forever and perhaps they are the layers of this universe of expanded atoms with their own gravity bouncing with the other dimensions also altered in those areas of the celestial vault and when exploding the layers get encrusted immediately and that way other ones further in parallel universes or in its defect the walls of this one with the other dimensions and those walls incrust themselves and starts to look like a point that swallows everything and compress it endlessly.

And in the past we believed they were just the rebound waves and the explosion reverse, but our author as he studied the alternated dimensions he thinks those forces could never be able to compress material eternally and is not a pull like pulling matter but on the contrary is a pull of pushing matter to a center by layers from this universe pushing through the inside from the outside because of being affected by the primary explosion of the star and perhaps they will never be able to really settle that's why it will swallow and compress everything that gets near it as we know even light itself.

On the other hand, something science left aside waiting for universe studious of the universe like John C. Robles clarify it or give them other angles to study it.

In this theme is more likely to be that way that a simple return pull since a pull of return would have an end in times of inertia and a compression push of rebound wouldn´t be like that and unfortunately is not going to be proved until tomorrow's astronomy get to the stars and that way we will know how it really is and we know that studying everything what this volume says you would see the walls of the universe in other ways and other angles so you can create your own theories.

We can say to those who haven´t studied astronomy that astronomy is very inconstant and that many will change unexplainable things like this phenomenon of the black holes believing that the returns pull would last so many millions of years but for those who studied more we believe is just an incentive this explanation since we know this happens but we don´t know how and why and studying the structures of the universe exposed in this book you will understand it better since otherwise there was this information and unfinished studies of what really makes matter react in a black hole since there are other theories that think it could be an abysm and not a reaction but the gravity swirls that creates and astronomic observations dictates it would be other more trustful things like a return pull only and today we tell you is the easier thing to be space walls compressing things for being discomposed with the great celestial vault in the explosion of a supernova and that way create the black hole and the endless compression of matter and perhaps it does have an end at the end of the universe where those layers are incrusted correctly in other ways that doesn´t compress matter that way.

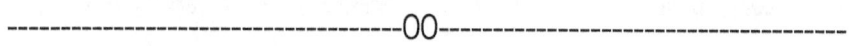

And so our author exposes in the subject of temporal astronomy And thus seeing universal time

EF23

UNIVERSES, WHAT ARE THEY?

Has we explained before, we can be in this universe because its own atoms are incrusted, perhaps is the speed of the electrons as they turn the polarity of the subatomic particles as they interact between them with alike values or perhaps is just a very hermetic division of the universe structures that we cannot even dominate since we are just starting to study them and at the same time is because just a few man have explained it and scientists and others have developed theories that lead to other fields that could end in this one of quantum physics if we get into that pad like the relativity theories that prove some experiments of the parallel universes and those who invented them dictate that several universes exist in the same space but in different time, but to really explain what it was it was difficult without the studies of the first ones as it would be difficult for the physics of tomorrow without this studies that we will define for you, the studies of parallel universes and alternate dimensions of the universes as we classified.

Returning to the theme of what they are, we want to go deeper and not stay only with the explanations we have exposed before since there could be and are doubts to clarify even doe they would have to be clarified by tomorrow science, we want to think about them and to be explained in their possibilities to save time to future astronomers or to guide them in themes that stimulate them in it, that's why we will give them explanations that we studied years ago like 3 or 4 decades; so we will expose how we started, that perhaps they are atoms that get incrusted between them and as others incrust themselves there is another universe and that's what practically what we exposed but also they couldn't be there just because and us here otherwise there would be other limits, just remember the black holes how they compress all that matter and it couldn't go unnoticed other universe just like that, but how a very important physic of the 20th century declared the perspectives of thinking about atoms dictate they are 90% pure emptiness and the rest is electromagnetism which means electricity and if we don't fit in a universe we could fall into free fall to millions of miles in universes until we fit or get incrusted automatically in one and the free fall of the atomic emptiness would stop there and the atomic unities would have to combine because of the turns the electron gives to the nucleus or things of atomic priority previously explained, but still we'll have to be skeptic and say, ok that should be visible perhaps or not just be that way, like some universes can exist and others won't incrust since we would see them fall very often perhaps or not, but existentiality maybe is not how we want and the edges of matter are so hard that they even create black holes that eat entire suns and doesn't break.

Then, where are they? Our author exposed they are there in other times and what is time, we previously explained it in other chapters and their dimensions and in this one we want to explain physically what things could be so other can clarify it when they dominate more matter.

So we will expose that the orange slices theory from our author, it dictates is like orange slices division and in every slice there is a universe or is a universe in a different time and where you can go to all the orange in every slice without crashing with the other slices and you can go from the rind to the center of the orange and every slice has subsequent divisions and those are the dimensions called Eta Carinae A1.2.3.4.5.6 and so on, or IC A1.2.3.4. etc. ok until there we're fine, but why is it that way? Well, if there are those differences of atomic charges that delimit and even today they cannot be seen, but we know that's the only way it could be and not otherwise, universes in the same space and different time, that's why we have the need to explain how it could be; well, there are several ways, one is primordial in quantities and atomic values as we said before and the others are values and structural atomic sub-values that we do not yet know, but before we go on we have to abbreviate in a simple way for our readers, that's the way the universe is, not how we thought it was and there is no way to see how really is but that's how it is and that's it and there are several universes near but we cannot go only with machines not yet invented but we know they would vary speeds of the atoms or would push with polarities or induced gravity in every angle and we'll appear in the next or other universes, we know it because of natural phenomenon and not identified objects have vanished before our eyes every year in radars or in videos and there are atoms and particles that vanish in front of us also but is not enough for objective discrimination to see that only but a series of things there are of physics experiments that prove it, but even so returning to the theme of what it is and where is that so rare since is immediately from here and just that but humans always want more explanations, there are other physic laws and that's it, so returning to what it is, they are perhaps walls of various atomic structures of atomic polarities different to our values from this universe and we will go there when we can break those walls from the limit of our universe and are the electrons but when we break it in a controlled way we could go since nature can't break it and we have to do it artificially maybe neutralizing the atom in its polarities or physic structures and where we fold it and fold it again since after all is just 90% space emptiness and the rest is electromagnetism but for that we have to study more and more until we make it since not even the black holes can only compress it to the extreme since this universe structures are very strong, we also know atoms different polarities from the different dimensions in its modality of positrons antiprotons and antineutrons change the structure by affecting the bases of what they are and that is because they vibrate varying speeds and create a different matter that disappear from this universe like other kind of matter that exist here in other time but we know it can be done due to thousands of declarations and anomalies and other experiments, we have explained it in other chapters from this book, but returning to the same theme, beside of what we have explained, the atom formed with protons, neutrons and electrons in their orbits and today we know there are antineutrons, antiprotons and antielectrons or positrons and like tens of subatomic particles that manifest and physics are cataloging, but to not just stay like that we will say that they perhaps are walls of different values as we said before and they are here where we are but we cannot see them, maybe nothing like this magnetism can be seen and everything could be of some king of magnetism or maybe everything is just like that been there, and when polarities combine with each other since everything is magnetism which means is not the 90% but some say is the 98% of pure emptiness and the rest is magnetism, imagine how many universes or dimensions can fit in the rest of the 97% more of the emptiness of the atom in their correlations, but even so is not just that, perhaps we can be there immediately without bothering this place and that's it, and perhaps there are other things some polarities that block us to go there and also divide if we go there since having other values of polarities that rule atomic charges of the electron there not just with the turns of the nucleus, but as such turns are created and other types of charges not yet studied or discovered by science or imagined other kind of atomic matter are created when the nucleus is affected or simply been that way with those differences or with other magnets polarities or to be more understandable we can call them Quartz or antineutrons and antiprotons positrons or failing that a very different matter electrically since the spinning of the electron, so you can understand how many variants there are, could affect the core and have another existential form like the bullets of the old airplanes in their machine guns, they never hit the propeller or simply different plastic or like the alternating current and direct current but in other senses that makes it impossible to see it from here and the most important, to exist in this universe since here would be the nothing or simply won't see it and the other part are just different charges from this universe that would create the extra dimensions from this same one and so that way there would be other polarities that make it seems impossible to go and also there could be other species of magnetism adhesives very strong so you can understand about this kind of polarities or other kinds of structures blocking and of limits and perhaps that 97% could be an amalgam and it must be of polarities and extreme complexity that creates parallel universes in the same space but in different time or polarity accommodations and as they combine with every polarity nothing can be seen and exist there or perhaps is impossible to go only isolating us from this universe and not in the same space but in another one, but it has to be explained that way since it could exist there immediately without bothering this one but with extremely different polarities and perhaps very simply different but impenetrable without neutralizing the atom or making worm holes so dreamed, or cavities to go there, and that's how it is the universe but only by going there we will be convinced about it since we know that's how it is because of physic evidences from others experiments and the complexity of the universe and by a series of discrimination of objectives we know they are there aside of this one, as well as dimensions as entire universes ready to be discovered when our technology dominate more the atom in its gravity forces and secrets of polarities.

On the other hand we know if we go we will open a world of sidereal travels and intergalactic transmissions that will give us more technology and knowledge that will take us to the era of light of humanity, and returning to the theme of what is it, we know being there is the only way to travel through the universe since otherwise we would crash the moment we leave the solar system and lets not say at light speeds but at simple speeds of normal cruisers like the space shuttle and it would have being impossible to get to the stars for us and for others if we didn't reach those dimensions and time abortions of time, for example if we don't create an artificial gravity bubble where time stops we couldn't go since it would take centuries to go anywhere and that way we would know we stopped time and also we would know when arriving that years had passed and our author says and firmly believes we will achieve to adjust that time also by other dimensions that others dominate and that way we will be able to travel through space time, by other ways it would be really impossible and that we have explained it in other previous chapters, but returning to what else could be those dimensions since you know we define and cataloged them in other chapters of this book the best we could, but also we know they could have formed because of several explosions of previous universes since there wasn't time in just one universe for so many elements according to the thoughts of the author and some kind of serpentines were created of layers from the previous explosions of this universe and as it went forward in its first steps by the universe in its expansion they were signed with different physics laws and different times for sure, but asides of polarity what other things are they? Someone must go to see it, but perhaps they are really strong and big layers of expanded atoms like this universe but in a rhythm or atomic time that place them that way but they are so strong to penetrate as it is this universe of going out and they are right there next to this universe let's say every foot or meter dimensions and in every mile a different universe and perhaps there are tens and hundreds of dimensions and we could create some day or travel to some that are ruled by other times and leaving and returning to it we could get out from this universe to 10 000 light years of distance by other artificial or natural dimension and then 10 000 years to get into this one, a few hours of difference only of our departure and that way we would return also and we could even create induced gravity effects that could stop time and forward it and only the universe go in the expansion, which means we could travel in a determine distance but we would have to adjust the watches and then look for the hearth and in what area we should be in the expansion and that's it, in fact we think there are civilizations that have achieved it and we have to achieve it too, and perhaps we will have to synchronize radios that are transmitted in temporal abortion and make molecules vibrate from other universes or their vortex and that way obtain great information from other civilizations, and that important is to advance in this science of astrophysics since we could have medicine and thousands of benefits, maybe.

That's why we have to think more in the atomic polarities and study more the atom and not just the light and astronomic optic since seeing those dimensions we would open earth to whole the universe sooner.

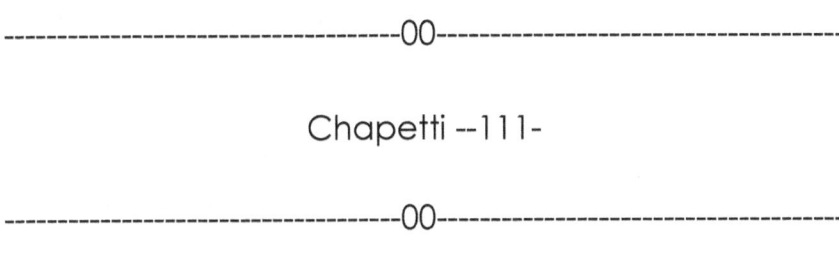

Chapetti --111-

----------------------------------00----------------------------------

And so our author exposes in the subject of temporal astronomy

EF24

ICAB4 TEMPORARY RECORDING

Universal temporary recording means how we have exposed that the things remained saved in dimensions and if someday scientists of the future travel in time, if they achieve to go back to the past, they will find the temporary definitions that the author invented and would be very helpful like the denominations that the author also invented of ICA1 F1 or ICA2 F 20 or in the past IC A1 P2 or ICA2 OP 4 what they mean in this dimension in the future or in this dimension ICA1 in the past and the dimensional categories and levels that they are able to achieve in their temporary trips even so or the stays could be of another form like the possibilities that they could be the only ones if you are more aseptic about where they could reach the past or future but even though please remember my readers that we will always be waiting that maybe the universe isn't like a simple celestial vault and that maybe there are other temporary stays and more complex that we will explain below and the previous ones like detailed alternate worlds that were built to alter the past and to follow the normal lines. We will also explain the reason why the past cannot be changed or the things to disappear like the woods that could have burnt out in different crono-nauts wars of a thousand near galaxies or maybe only meteorologist phenomena's like meteorites that would burn the woods in the last 50 million years to the earth. It could not be that traveling to the universe it could be static in the same time and many complex things and weird could happen like going back to the past and that it would not be the same, but this future yes to continue or maybe yes to be the same but only in ICA1 and the rest to be in parallel diverse worlds that would occur in natural circumstances and of the complexity of the celestial vault or created by other beings maybe for the study but we know that for sure nothing changed in 50 million years so that they can have a reference before you.

Catalog with patience in 10 days the dimensions and extra graphics and all of these in 10 occasions on

Saturdays in the morning and we could say that we live in a hard drive, for sure electromagnetic, since it has all the qualities of being; the celestial vault is full of expanded weird and complex atoms and besides it is ejected in speeds without stopping until an unreachable end. It enclosures suns and galaxies and even black holes and areas like the great attractor and it sustains without a limit the different factors of plasma matter or weird temporary horizons and everything that we might imagined it could have. On the other hand, recordings of different cloned atoms by the expansion and its speed while going forward every minute and that is the celestial vault of pure rays and gravitations extreme waves of atoms expanded and the evidences of all what has been previously explained it really could be a hard drive. The theories of parallel worlds like worlds like if the events were others and like if stop would mean to continue and so forth like if saying to continue meant to stop in the same planet and existing parallel since in the past it continue everything the same until the present but while traveling in dimensions we know that if someone changed things or killed somebody it would keep existing in all generations until a million years without changing it but to go to the dimension ICAB 2 or ICA D4 there we could find those alternate worlds right here on earth existing at the same time, we could say that is not possible but it could be due to all what we have explained in this book.

Even like that it could be that those dimensions are there and temporary differences to live with other people or only to travel through the universe. On the opposite it would be impractical to travel but nothing changes in our eyes and nothing disappears that is why what existed one time it cannot be changed since it created an endless series and when a crono-naut arrives to the past to change something that has already happened, once a series of events has been created and nothing happens. Besides it could not be that if they change something and it is the same that they exist, right?

And so our author exposes in the subject of temporal astronomy

EF25

FUTURE PAST AND THE TEMPORARY RECORDING AND THENS OF TIME FUTURE ICA1B2-F AND ICAB2-P

That is why the future then is located within parts in the celestial vault of dimensions extraordinary adequate where there exists thousands and millions of various times, both in millions of years and as millions of years pass into the future in capacities of the hyper universe that we have like a home or celestial vault and that I believe the big exploration and questions is really like this. It could be to have all the times at the same time or maybe they will begin to open in a limit since in another way it would be a lot and for the universe age and earth itself it is nothing a million years since we would need to see if we could accommodate everything in 100 millions of year in the future or 50 millions of years in the past if in 13,000 thousand millions of years that they have the bing bang calculated and we could even speculate that the dimensions of time are forging every millions of years of different subsequences and maybe they are more and maybe they are already like we say in the final universes. All the universes found in different times one first basic of 800 thousand millions of years of 400 and 300 and 100,000 the end of this like in 200,000 millions of years but for studying time and understand the basic we will deduct that we are what governs and our time is the present perfect and from there comes out all the time of the universe and then to understand everything we will know and we will transfer to the angle that that is not like that and that we have to multiply everything maybe for the thousands of millions of years more.

So without multiplying we would have seen but still being in there we dare to say that maybe in the end the future is there in that area or fit the immediate futures maybe 50 through 100 millions of years. Or maybe you believe that the dimensions are traveling with the expansion and leave something here of celestial vault or over there it will be the celestial vaults that created all those times?

Well from there we will guide ourselves to see what they are and where they are since today only 13 thousand millions of years have passed and the other future is in the expansion towards the expansion of the future and no celestial vault is created unless the previous ones have stopped and are here some old in other times before to all of this. It is good to render it to understand time more profound since we are talking about dimensions and parallel universes and we have to understand physically where they are. Well, the future times we will study like in 50 to 100 millions of years and the past times in the years of this universe and the ones before also at the end and most likely at the beginning to delve into this.

But in this chapter we will see how the capacities are and we will classify them physically like extra added to the previous chapter that explains how they are over there towards the expansion or above the future and towards the bottom of the past or in its defect towards the end but we will explain only this universe. For this we will tell you that it started with the explosion and immediately formed thousands of time and dimension and each fact that is what we want to explain, and what each object is, it will remain recorded in the celestial vault explained previously because it could be like that and form that point we go to the next.

Like the one that is there in microseconds more and is there in 50 millions of years more and 1,000 simple years more in our explanations but where in dimensions that there are in different times, like we explained in previous chapters, and such dimensions belonged to that celestial vault and are accommodated like dominoes, one after the other and the past already stayed all the same and saved in a form that even our studies don´t explain everything like they are accommodated there and remained in its defect that are traveling with the universe in expansion. Or they remained out of the expansion and are in parts of extra dimensions that don´t affect the actual expansion if we see everything lineal. Everything must be to understand it in our world or we would have to make calculations of thousands of atomic molecular lace of antimatter to see where they are.

On the other hand, we expose that in the future others will do it and will define like this the universe of time in the physics of tomorrow since in this volume we offer only where it could be and until it is proven it would be like that. Only one past to understand other things that explain the book it makes reference.

We also expose that they will be like we say in a stair to the future 1 it would be 1 year ICAB 2 and to the past it would be ICBP2 that they meant to be reference of the galaxy ic1101 in straight line to earth dimension A one step to future B, in other words, it is first B and the denomination F that means future and so on and then with decimals at the end of the alphabet until all the years to the future and microseconds to the infinitum. The other is ICAB2P that means IC reference of the galaxy ic1101 to the earth A1 that is the dimension of this A1 b that is subsequent time p2 that is the following past. In P2, p is the past and so on to the beginning of the universe since the past it is there but we could also see better only 50 millions of years but to not complicate ourselves we will do it sometimes with 10 thousand years of references only. That is how it will be catalogued and there they are dimensions of what already happened and it was saved in temporary dimensional capacities and the gravitational waves of the universe or the ones that are there when you jump for other dimensions in time and that they are not gravitational but those for others more aseptic.

THE THINGS OF THE FUTURE AND THE PAST EVERY MICRO SECOND TO THE INFINITE

-----------------------------------00-------------------------------------

CHAPTER 6

And so our author exposes in the
subject of temporal astronomy

-----------------------------------00------------------------------------

And thus seeing universal time

EF 27

The Orange Segments

As we know and believe faithfully, in the theory of orange,
since otherwise it could not be that there were parallel
universes in the same space but at different times.

In the past, several scientists decreed that parallel
universes existed but in different times and to think how
could be found because the only form is the theory of
orange, as they have to be here but spliced. That is how

it could only be and in reflecting, our author explains that in the way that the theory of the orange segments is that the universe is formed by dimensionless segments that divide one the dimensions and other universes and that they are here but in different times. Explained in the same theory of the orange to be verified scientifically, exposed in previous chapters and previous books of the same author but even so we will detail them for his approval and that is that we believe it is necessary for his greater understanding and that what is next like the beginning of this universe in the primary explosions of the universe this formed several dimensions in each universe, since we believe there were several universes before this we are of that thinking as our author exposes throughout this book and before this universe other explosions formed at least 7 to 15 previous universes. Besides with long distance and each of them with different dimensions or one in its defect the elements that were formed apart to not delve into this as it is already exposed in the other chapters.

We have been exposing that the universe as it exploded formed various segments of an orange and that each one of them has different dimensions and each segment is a universe. The earlier universes created several dimensions apart or the possibility of going to them when entering other atomic electronic charges because they are surely formed by this difference by different atomic charges and things fit into each charge only if the walls of each dimension are very strong and even withstand atomic bombs and cataclysms of explosions of super nova and first create black holes that give and break because the elements of which they are made are like springs of steel that will never yield unless they are founded with an atomic flux. Like if it would be to neutralize or polarize the charges that form the matter and we have not yet achieved that technology to dismantle the atom or atoms and travel to other dimensions or universes or create the so-called worm holes artificially and that is the reason we cannot go over there but science has theories of how to do it and even alludes that alien technologies would have demonstrated to scientists with secret experiments as which we all see today in science fiction movies.

That is the reason we dare to say that the universe created various universe layers creating segments of orange since the beginning until the end and that each one is a different universe. Maybe in comprehension and others in expansion but besides we don´t know if they have already stopped existing or just go back and are one in top of the other during this times. Or if they are next to one another in segments since they could be the dimensions and the other universes which are like segments of the past and therefore, an interior to this universe.

Each one of them with different dimensions and having various segments on which each is a different universe and apart has various individual segments and each is a separate atomic dimension like we alluded on the previous chapters. That we see the atom and it is pure empty like 90 % and even though if it wasn't like that one would fit atomically with those levels of charge that would drop down on a fall without an end to an empty until the universe with our charges fitted on the atoms of our charges of the matter limits, which would maybe be the velocity of the electrons as it turned the atoms and the different polarities that would be in them like charges would be like keys in them or maybe subatomic charges that would be discovered tomorrow. By then be an existing strange alien here without seeing it in temporary and dimensional walls.

Therefore, each segment is different and each dimensional segment is the same to the ones created artificially by the spaceships that we would travel on the universe in the future on which they would depend, first of all, or they would crash with the first rock that would cross on their way. This was something that our author confessed to have always thought of but that the scientists didn't or at least not during those days until it was specified to be.

That is why the universal segments function like that and that is how you would be able to go to the end of the universe on which it is represented as an orange and the skin would be the end or the most far away to go to the universe from the top to the bottom and towards the end or to the beginning that is why we would talk in another occasion. Today we are exposing what would be of each segment and what would happen as you go from one to the other. You could possibly go through the entire orange starting from the skin and going to the center and to the end and go around the whole limit and without crashing with the other segment that would only be there but at a different time. And that as you went to each segment it would be the same thing and you could go to an entire universal orange the same way without ever crashing with the other segment. That way in each of the dimensions of the segments created artificially pushing the matter aside and the rules of this universe temporary speaking or natural which there has to be and classified by us statistically since everything would remain just at a different time.

That is what means to say that universes existed in the same space in different times. Then in these same segments there are a lot of things and different phenomena of gravity and we believe on other atomic structures which vary and form multiple universes and the entire celestial vault of the space and that is what the theory of the orange is.

So if you arrive to the past and burn a tree within seconds it would resemble and every microsecond. to the infinite.,

----------------------------------00----------------------------------

And so our author exposes in the subject of temporal astronomy

EF28

PRESENT PERFECT AND VARIANTS
PRESENT INITIAL PERFECT

Then we dare to declare that statistically it could be the perfect present, it is possible to be at birth the first ancestor since there will be the root of our genes or the microsecond to the infinitum which is before the temporary recording of the celestial vault or for the skeptics of these recording ideas it would be the microsecond before creating all then default to temporary domain. We would have to arrive before nature itself and maybe it is a weird protected place with layers of the universe that hide from the rest of the creation and protects with quotes or turning electrons in rare positions and that it is there, we will pass by there but only the protagonists of history of our lives, by defect and maybe other beings from other planets and times will not visit us. Maybe they are affected by the waves and layers of dimensions and it might be difficult to get there. If it is not at this time and to travel on the universe you would need to evade time and besides for sure it comes from other galactic times or universals because not necessarily came out of the hour from here to eat, but in their hours and while evading time to save the trip they dressed up as of this time and if their star is at this time or their galaxy, the stars rule the gravity and the time apart. It is believed that each star has a different time and each galaxy on its variants that have not been studied yet. Maybe their temporary adjustments need them for other things for which they came to these areas of the universe. Maybe they don't come for what we think like maybe to see the star or adjust propulsion or something like that and not to intervene with the species human host in this case since they would be needing to enter the present perfect to really be with us. Maybe a machine or equipment is needed that consume much energy and it could be strong to do that or maybe besides to create problems to do it. Since it would affect our future and only if they are linked with our history it could be seen without any problems. We are not saying that it is the case bit it could be for some of them since with any change of traveling into time they would be separated of this tie and adjusting to go back to the present. To talk what we have to offer is not attractive and maybe only for others in the future it is more like us. Maybe if there are not prohibitions of quarantines or political situations that have the planet blocked or in its defect, that they only be technical things that cost time and energy since they appear, do not notify and disappear with no trail. Before cameras and radars many years nowadays and maybe apart be the ones that adjusted the present perfect on which there could be in this dimension and could be here. For this it would have to be very important what they need to talk and even there nothing would change because the future can complain that the human race wasn't like it was really supposed to be due to their fault. So maybe that is why they would only change good things or maybe nothing.

On the other hand, we think that there must be several or many variants of the present perfect but only there could a paradox be created and alter the future from the past and that we believe would not disintegrate anything in the present or future or we would see many things disappear and we do not see anything to disappear before our eyes and that is proof that the past cannot be changed.

Then if you arrived in the past and burn a tree it would appear in seconds and every microsecond. That is because there are many civilizations o maybe only natural events that could have changed since we travel on the great celestial vault towards and expansion on which there could be natural phenomena that 50 million of years could have created a paradox and woods could get disappeared, as well as animals and that never happens here. With this we complement what could be the present perfect that is before history itself could reach it and get here, and to be the minute before recording the next one or to create the then of that future. That is where the present perfect will be, that is the basis of why it is not possible to vary history with things that are not connected with the present perfect itself.

-----------------------------------00------------------------------------

And so our author exposes in the subject of temporal astronomy

EF29

THE THINGS AND TEMPORARY RECORDS
OF THE SUBJECT OF THE UNIVERSE 3

The then things are the past and the future, right? We have talked a lot on this book about a temporary record that might happen, be and exist like it could be but to think skeptical we have to adhere something extra rather than the story of tomorrow and yesterday. Even though we have detailed this in a few lines we still have to look into this. It would be like a record and we brought it so close that it would then be to see a record which would be the same since you think a record is made out of particles and engraved atoms in a magnetic tape. But what happens if it would only then be every microsecond or thousand of a second to where they went or why it would be traveled over there? Since maybe it would be through the dimensions that our author exposes in other chapters ICA1 and ICZ64 or in its defect the variants that could have fallen artificially getting lightning of gravity. That is how it should go through other dimensions and universes and now that so far it has not gone or maybe just a type of atomic speed but maybe it is just that in the past that is how it happened and the universal present perfect s are not our lives but other physic laws of the universe about to study. That is why until we know them we will understand better those present imperfects and present perfects to understand the past. The non perfect and the future perfects and not perfect since going once it has happened it would be the way to prove the trips through defined dimensions that are different, atomically and spliced with the polarities that create maybe through the existence while giving other existential values to the nuclear atom and therefore create other times and not only without fan blades interconnected in three speeds. It would be there to jump through dimensions ICB F6 maybe ICAB F30 and for the past it would be ICAP 3 or ICAH 30 and that they mean in a straight line to the galaxy IC1109 of reference the alternate line a, b, c, d, or h, e and with variant 36 or 30 times of remoteness of that straight line in differences of charges in time or polarity differences of subatomic particles that create variants of time and the dimensions as you see in laws not written yet by physics.

And that quies were formed in layers at the beginning of the universe by exploiting the expansion

Or in the way of this expansion. And so go and understand these then

That is the past and the future in textual words but even like that it is not difficult or easy to understand. The opposite, we would not be able to travel over there but what comes next is difficult to understand but the same it is contradictory according to criteria and experiments of tomorrow but maybe it is easy as I wrote it before or in its defect be mixed.

Maybe it is a multi-universe hyper complex of everything created because of the same time and everything is already written on the existential universe when starting in micro seconds to the infinitum. Or better worst on the millions of seconds to the infinitum and for each second exists a human that will travel in time towards eternity and the most long lasting objects but literally we know that the oldest objects like the galaxies and sun die and will be over one day but what they left as atoms and civilizations are over there. In this time if we were to go to the future or in its defect only would be records of matter according to extended atoms of different celestial dimensions and of gravity rays that could hold a record of objects in areas of the expansion of the universe. Everything will be copied and recorded in a giant disc of the universe that records all the existence and that is how in its defect it creates the famous things of the future and the steps since in any way if you travel up to there it is either one, right?

That is why we repeat, decreed and dare to say that the things are temporary records in green celestials created by extremely expanded elements of helium hydrogen and other elements with gravitational rays and dimensional walls of the hydrogen and helium or expanded in its expansion or as existing creating such gravities of the matter and of the space vacuum and the matter of the dimensions of the universe or are then saved in alternate dimensions each micro second straying and accommodating like a domino hand in layers without an end into the future in the universal expansion and where that phenomena can be created that is maybe in those limits of layer creation in the expansion of the universe would be better appreciated and which we call future past dimensions.

Then you could go to the then of yesterday only for a dimension to create an artificial dimension and the skeptical would say that the world of temporary record doesn't exist and that the layers are just a trip to the past by the shortcuts of the universe or by weird universes for where it could be traveled back then. Once you arrive to them but nevertheless the rules of junior dictate that nothing can be changed going more and I would have to arrive before in present perfects and even like that change millions of hours or years, otherwise you would see before their eyes change things – physically like burned forests or persons that disappear and appear and also disappear a lot of things each moment that someone changed the past, natural events for traveling in the universal expansion. This would have happened in 50 millions of years and like we all know nothing changes and those material things are there. Buildings, forests and things and nothing changes. Only fortune tellers change things but from here seeing the tomorrow and future time, not traveling over there and they would be connected to this present perfect like lines that have already been written. Also parts of yesterday, today and tomorrow.

Since nothing has ever been changed or nothing has been disappeared before your eyes, no buildings, no personas and no forests, it is a life of 80 years and it is the proof that it is how it is but it will take decades and maybe hundreds of years until we are there and then us humanity as crono-nauts and that is only how we would know when our sounds will arrive to the truths of the science of tomorrow. That is why we do an emphasis that maybe biblical writings are real like the one that says that everything is already written.

Well that is how all the information of theoretic scientists of the quantum physics would be formed and how it is still not understood or it is not explained good and today we are given an explanation of logic deductions for single target discrimination and interesting of all science or thesis that will start to appear in the measure that there is more information of academies and diverse knowledge of different books and diverse authors of astronomy and physics. As others like our author that have special particular studies of astronomy and geopolitics and economy. People like him use to write with a lot of cultural information of the modern era so others can take advantage of those cultural and theoric expositions that start with angels to advance and to begin in them would have taken the same that these authors like it took our author 40 years of studies of times and parallel universes.

People that studied and will study and will come out to literature and that I think changed the way of studying like we have said in the era of the modern communications.

-----------------------------------OO------------------------------------

CHAPTER 7

And so our author exposes in the subject of temporal astronomy And thus seeing universal time

EF30

THE GREAT ATTRACTOR AND OUR CHALLENGES

We know that many mysteries of the universe will remain in the future to be resolved and there might be millions and not tenths or hundredths since other 7 universes were created before and even other weird things. For example, the attractor that is real and the so cruel universe that there are millions of civilizations dying and only a few advance are saved. How discriminative the universe is and what dangers we have like the great attractor that could be devouring entire galaxies and killing millions of civilizations even the most advanced ones. Let us know don´t thing about civilizations but of all the animals that will be dying there without us really knowing with exactness what our destination for the future is. If we don't prepare ourselves like human species and we arrive to the spacial times and dominate the matter quickly. As a civilization we could survive without technology and how much time before attackers and catastrophes would arrive like that of the great attractor situated at about 250 millions of light years from here. It is speculated that there are parts in which the galaxies and matter are being attracted 29 millions of kilometers for now, something that it could not even be imagined. Also that they would be devouring entire big galaxies and everything that is 150 millions of light years would be consumed if it is not removed from the expansion path of the universe. It is even speculated that maybe millions of civilizations have died with animals and human beings without anyone being able to do something about it. Due to that swirl of the universe that it cannot even be faced since it is situated at about 250 millions of light years and our galaxy escapes to that tragedy but what other things does tomorrow will bring to us like a civilization. All of this makes us think that maybe we are living in a prohibited era for the visitors of other worlds of ecological beings. Maybe we are isolated due to the galactic wars between super civilizations or gods of other times and galaxies. Even like that we know that we need to prepare ourselves to be saved like a civilization or on the contrary we will die. Maybe everything that we have forged will stop due to the missing technology. That is why it is urgent for us to study more the universe and not be at the margin of the elements like those civilizations are which are being devoured by that attractor that could be a gigantic black hole that attracts hundreds of galaxies and that since these points of the universe we are unable to see anything except for the strange trip of attraction which are living everyone that is near that area and that is being devoured and it is disappearing in that point. We ask ourselves how sure we are of how much natural danger we are facing in the future and of how many human beings there could be in the universe and how dangerous it would really be.

Maybe if we had a radio of temporary transmission and dimensional we could know things and could hear millions of exclamations and crying of entire civilizations and we could prepare ourselves better.

Of other civilizations if this area is protected by gods or other human beings and created a species of quarantine how much time would it last and how sure it would be to be isolated from other technologies like the ones that might the victims of the great attractor need. Maybe it is the comprehension of artificial spaces of time and to go out of the danger are or travel teleporting to other dimension and areas of the universe.

------------------------------------OO------------------------------------

And thus seeing universal time

EF31

EE F 1 THE RADIO OF TOMORROW

And so in the dawn of the 21st century we deducted that humanity, in other words men, could communicate to the exterior space.

Theories of Juan C Robles

Num 6

Deducting the forms of communication to transmit extra planetary, extra galactic and extraterrestrial the applied modern technology should follow my suggestions of object discrimination and doing a summary of what frequency they could use the ones that would reach the planets and near parts of the earth. Inclusive far from the earth there could only be the following by logical deduction, the better to communicate with its parts in other planets within light years of distance without having to wait 100 to 1000 million years. It would be sub-spacial with spacial time adjustments to abort the time and adjusted it while arriving and exiting as explained in previous episodes of how the time would be adjusted and which we recommend to read.

That is why we deduct that all radio that wants to transmit towards the stars have to be the most viable for such objectives. That is the reason that we think that it could be an objective discrimination something similar to traveling over the universe, evading space time like a kind of bubble of temporary abortion or stimulation of particles that while they travel within the universe they would avast time like a king of flying saucers and achieve this way light years in hours or minutes or days. Otherwise, it would not be practical.

And we believe that for this it would have to be an inter-dimensional transmission or that it would evade the time space so that it would arrive so far but would be impractical because it would take hundreds or millennia and for that reason we believe that the stations of monitoring of galactic transmissions are badly focused and they have never thought in it. But we believe that first they have to listen and try to tune what they could be transmitting the beings or inhabitants of other worlds and it could be to listen through tuning first frequencies that can evade the space time and where they could be those frequencies. And second the frequencies close to the dimensions following ours as ETA Carinae point a1 to point a2 or in its vortex or in its borders that is the following dimension after the first that is where we are. They could be in the way of adjusting times frequencies sent passable from one place to another of the universe.

On the other hand, nowadays when studying the radio the author believes that they should be what is near to evade the time space, otherwise they would not arrive and for that reason you have to listen to the vibrations in the molecules of the atom or in the atoms in the subatomic areas where they may be those frequencies that must be, where atomically it must be, or in the vortex between each dimension of each universe. Read ETA dimensions Carinae point a1 in the book to learn of possible dimensions and vortex between dimensions since only there or in the strict studies of modern science you will be able to hear maybe galactic transmissions. And perhaps it is where certain physicists know that they could be passing from other dimensions or from the vortex of each universe and dimensions to those transmissions. There is perhaps this dimension in the subatomic particles had not been studied well today but soon it could be possible to access that technology when the communications begin to experiment there

And it would be the era of the light of science perhaps to do so would be the most extraordinary database of communications in this world, since humans could get from medicines to fascinating technologies that would help humanity. Well, this day I encourage you to go in the right paths for future astronomers and future physicists who have not yet studied or have not been encouraged to deduce that parts of the atom are closer to the DIMENSIONAL BORDERS TO ACHIEVE THOE SOUNDS AND TUNES maybe only subatomic boundaries of the already discovered particles. Where other universes and dimensions would pass through and there would be vibrations with messages from other planets. Even so it may be necessary to access subspace bubbles being transmissions created in time evolution and we have to study how they could be and for sure there might be messages of the universe and we also have to study how we could accelerate those studies in this regard. We might need to create a worm hole to get there and access but maybe experimental laboratories of electronics achieve without those holes and access those coveted vibrations of frequencies of cosmic radios with extraterrestrial languages that maybe one day say "Hi" while accessing them and we need to know how to change them in time. Maybe they have been encrypted for traveling into time and we might have to understand better the time by studying it like we do with books like this one, right?

Since so far no one has observed there since physics barely knows the space time and barely they are beginning to define in theories of how they could be located those dimensions that like Albert Einstein said occupy the same space but different time. Where at the end and within the book are also explained more scientifically all the definitions and broadens a spectrum to study the sub spacial time.

Even like that we have to investigate which are the real borders of the matter between each dimension and if those extraterrestrial beings make the atoms vibrate the vortex of the dimensions or of the universes or in its defect the transmission has an effect of vacuum space time in its molecular vibration and that is what the galactic radio of the future could be. And it will take so that the scientists achieve that finding but as I explained to you by objective discrimination only like that is by space time absorption in its effects of how they could transmit and tune in galactic radios.

Since there they could be trying to communicate to their home the inhabitants of other worlds from planets to planets and not in normal radio frequencies that would take centuries to arrive by the molecules or atoms of the universe of the normal expanded space vacuum.

Therefore they may be in the stimulation of neutrinos Antineutrinos B2 positrons or anti protons or maybe only other subatomic particles or charges between those particles hat manage to perceive the vibrations of vortex transmissions and near dimensions in another way so that we have the worm holes or dimensional jumps we could possible know. ....THE END...

---------------------------------OO---------------------------------------

EF31.2

EPILOGUE OR INTRODUCTION
THE ADVANTAGES OF SPACE TIME AND REAL TIME

Therefore by this means, we expose that the theme of this book "The Time and its Dimensions and Location" or even better that that time spaces will always help all humanity and not only the astronauts and physics to conceive better the same space but will really become the BOOM of the human preservation. In fact, it is better for mankind to exist like we have exposed in this book since out of the big misfortunes that will bring to the population and the lack of food and practically abandonment of other beings or quarantines that belong to this world named earth or that in its defect the fact of being in the free will or literally alone in this space of thousands of galaxies and solar all around de exterior space floating from a water and dirt sphere, we know that all misadventures, challenges and catastrophes could be desperate and immediate could make life on earth disappear or most likely humans. It would be the ones that the ozone layers would ran out and that is why 21% of oxygen would degrade of the common air and 78% of nitrogen of that same common air that is only what we inhale. To degrade the plantum layers of the oceans, which are responsible of the 80% of the natural production of oxygen, that is what we inhale of that form and the earth would be in danger.

We also know that it would threaten life immediately a meteorite of impact in which ever ocean of the earth causing a big wave of thousands of miles of height and inclusive an invasion of extraterrestrial beings, slaves or predators of space and that would be the catastrophes that could destroy the future of humanity.

And let's not say to have taken to the planet to some nuclear wars but that is politics that could be changed and that we corrected but the rest of the dangers nobody could possible change them and we will have to deal without technology of dimensional time of the future, especially the ones from space time to be able to persevere the human life and that way be able to save entire cities in domes of other dimensions or most likely dimensional temporary domes and create an unending dimensional barrier against extraterrestrial invasions and natural elements liker filtering and cool the rays of the sun with barriers of space time without mentioning the many benefits to science like the transportation and dimensional teletransportation, the sideway trips within the space time that will give us the chance to leave our times and travel securely without crashing with an asteroid or big rocks and high speeds on the universe. The teletransportation of elements for its cold function and to achieve wonderful materials and thousands of advances that will bring the comprehension of time and atoms and this way face the cosmos and its dangers and inclusive achieve dimensional galactic transmissions to obtain medicines and technology of other civilizations while creating transmissions that stimulate the atoms of others universes or in its defect to create transmissions of evasion of time reaching out thousands of million light years towards other parts evading time in seconds and travel in an hour and arrive to adjust the time and evade when leaving this space time let's say at 9 a.m. and arrive at 9:15 a.m.

Or what all the scientists dreamt for decades to transmit to the center of the galaxy or whatever place of the universe evading time and being answered in seconds at light velocity but really should it be light speed? Evading time and stopping 100% of its advance to the passengers or the transmission and to leave the laws of physics of this temporary universe and leave practically at breakfast time to thousand light years and arrive at dinner time and arrive to the Andromeda galaxy at 2 millions of light years in 10 hour and arrive 10 hours back. This we will only achieve with the dominion of space time, that is why the students of quantum physics and astrophysics and all the general public that like this subjects it stimulates to continue studying this themes and with the angles of this book to arrive at the limits of our scope of this matter and help grow and make our civilization more secured adequately since all those benefits will bring us to study the dimensional time and the space and angles that we give away in this book for its reading. They will save those 40 years of studies and time, hours of reading which our author dedicated to understand the space time and comprehend and know what could be. Also the best of where those times could be located, spaces and dimensions and all parts and alternate dimensions and parallel universes that could be and they can ensure that with your studies to start with this book or the angles that they will give this book extra and catapult science and at the same time humanity and will ensure it.

But even like that our planet has a lot of challenges and goals to meet since the economic deficits created for the lack of continuity of the monetary system make each day the population to grow and the democratic economy will come to the end of the primary stages to solve everything and due to the demographic explosion and the overpopulation they will make it to be uncontainable, thousands of budgetary deficits and unemployment. Everything will be focused to primary things and will not give opportunity to budgets for science like it already happens for technology. There will only be the way that it is convenient for them and besides not only to urgent things but they will be first in the national defenses of each country and the budgets of science will be vetoed and they will never have enough to be able to achieve dominion of the atoms and time and as we know the problems that will come will be difficult without a healthy economy of subsidies and with no corruption. That is why the world of dollars will be a hope that they will have to give to the overpopulation and the failures of constant finances like for the industry that they need in their month and even dead years of sales to be maintained more each time.

EF 32

THE FUTURE ECONOMI PART 3.3

That's why we know we would need this to achieve our goals has humans of surviving to a universe in chaos and destruction like previously exposed and quasars of gamma rays and many cosmic cataclysms there would be in our way to the greatest odyssey of human kind, the conquer of outer space and the universe.

FUTURES ECONOMY Part 3

EF-1

Futures economy as we know is a kind of special reserves of decree of the countries that in the future will change everything we know as society, perhaps is largest reach of our author John C. Robles although he has not been recognized but he has two books with copyrights.

Today we will delve deeper into that form of economy of tomorrow, and because he wrote the first articles about it he practically invented it and designed of how it could be and because of the next we expose the famous world dollars of tomorrow, they will come in a bad time of humanity with an enormous demographic explosion and impossible deficits already in 198 countries and very bad in the other 28, some, the most industrialized will come with the authorization of the world bank and united nations in the international monetary fund and other Asiatic organizations, authorities of that time and they were approved, we calculate or our author calculates in 2040 to 2050 where after decades of studies will be authorized the special reserves of humanity decree and at the same time will create the world dollars or free cash and that way will be approved and perhaps will vary the way to approve but it will have to be under the tutelage of the united nations in organizations created at that time for its regulation and that the economy doesn't prevent it to be and to reach it since the conservators will get scare and will believe myths about the hand work and other production instruments, but being proved will not be more consensus but to prove it and doing so humanity will come out of the Jurassic times in which has been for millenniums since future economy will come to stay forever and will be the propeller to catapult the future as never has been before, such economy is based in special reserves without any cost to the countries and will subsidize all the deficits someday, unpayable deficits that suffocate the countries and create poverty.

Such reserves previously explained will be prepared according to the interests of their representations in the United Nations urge and perhaps some will go to vital parts at the beginning and others don't, but some will be very good applied in parts and money packages that will relieve the nations expenses that accredit them and where United Nations dictates, thing that will not be badly received at that time, also there would be less differences between nations in alternate discussions, but even so everything that is free for anyone is welcome since it means less expense for themselves in many things and all the countries will benefit and those benefits of those world dollars will be felt throughout the world crossing frontiers without limits in all the planet.

And so in the end the world dollars will be approved for all and perhaps the controversies will be famous and the times vary in their approval but once they arrive that is infallible because the economies of the countries need them as the energetic to keep walking and the world dollars Famous in the minds of humans will not stop being and stop the hunger of humanity and the misfortunes of boxed and finished budgets and poorly focused budgets of our world but rather is to face the lack of tutelage rather than having created them before the division of nations in the 20th century and the previous centuries but these forms of reserves will be approved for being the only thing that could in democracy and other forms of government actually rectify the deficits of the budgets of the valuation of total production in the beginning and of the supply of value added to the global economy in its requirements of the budgets of the valuation of the costs of real economic existence in the real facts of the civilized world today for the world population in its demographic explosion and in its geographical demands and democratic and in turn in their percentages of governability and growth.

Because the economy of normal reserves failed and remained behind and only worked for some time and there are thousands of deficits and deficiencies that create very harmful phenomena such as not being able to go more in growth and at the same time provide the needs of the demands of democracy and civilized world of today let's not say the time of its approval perhaps of 10 to 15 billion human beings at that time even so today are already a real demand for thousands of things and will be an effect of ultra-capitalism or democracy without limits as they cannot boast of being very well without these forms of money any system that prevails because in any way the population grows and needs with it and on the other hand the automation grows and one day perhaps in decades or even centuries will create unemployment that phenomenon and apart from a growing world in exigencies requires a strong and dynamic economy that there is no or that the existing reaches its limits every decade at very high costs because in the great and better democracies there are low living standards and deficiencies has a lack of education and worse still very high demands of the population of jobs that there will not have and this form of economy alleviate all this little by little to the limit of providing the individual with the smallest free necessities while finding work or more still in the centuries that come give you the necessary for life if the work does not exist more time because apart from automation and robots and demographic explosions will end all types of jobs and we will be in an era of leisure of humanity and with these forms of subsidy reserves are adequately regulated.

On the other hand the world dollars once approved with most of the strongest countries and perhaps zero against the weakest will let themselves see and feel these new forms of reserves and will be like a respite to the deficits of the democratic economy that is already missing today and so there is no good and between ages of dreams of economists of the past as the perfect economy and perfect competition dreamed economists we see in all the books of today's economy in western universities and higher studies since the 1970s in all universities around the world as the perfect competitiveness economy of Adam Smith and other great economics.

And even these world dollars are the only thing that could make reality in a world of 7500 million humans maybe 13 thousand at that time and 198 countries as the earth is and then more and perhaps much more for that dreams of the perfect economy will be closer to being, when the world dollars are authorized and will be so.

As they are government decree reserves apart to create world dollars and inject subsidies to the unpayable and problems like Greece today or the lack of aid to sick or the debts of banks that finance subsidies to big banks so that they continue walking better and things like that.

That is why we know that the subsidies will come in bulk and in areas never seen and normal areas, the normal will come first and in the form of subsidies to agriculture and jobs and companies that guarantee jobs and benefits to humanity in each country according to the credits to be credible and that the problems they have since they could not be denied more than non-aligned countries to this great world order of economic truth.

And so they will be in the form of subsidies or better explained of free money for any country in trouble and there will also be personal for the normal people who are most affected by the failures of the economy, but first of all in those areas of workers that is very marginalized there will be where there are great needs perhaps hunger or war or maybe we know that there will be no way to absent the workforce that will be in charge of very responsible people who dictate which countries and who and which companies will be credible to those credits and subsidies and will not benefit more than those they dictate but it will be an international context very well organized fortunately and badly but in the end they will be certain and thus there will be monies and subsidies that will not be exactly where humanity is needed but they stimulate to other subsidies later and to get used to them and also what was spent on it will disappear and cease to be a burden for the countries.

Since what prevented them may be conservatism and will be defeated little by little but once they go, they will have no end.

For this reason they may be used instead of hunger in unlimited subsidies for research and science such as space and medical research that will also greatly benefit by avoiding those expenses for the countries and thus will be channeled a series of cash cascade to other areas much needed for normal people and countries in general.

And upon reaching those areas of research you will be able to find fresh resources from many and channeling billions of dollars for it and it will be a treasure never seen and unlimited medical research to cure the most possible diseases for which no one will be in disagreement and man's attempts to prolong longevity stimulate different parts that in turn late that soon will make them reach other areas much needed for what will always be there.

For each bank each factory and every country that needs them and justify having them will also be there to make the most of each day and thus change the horizons of the planet.

Imagine propulsion research to travel the universe and obtain medicines and techniques from other civilizations and mining resources in gold or other minerals.

Imagine 60-year loans to buy a car or 100-year loans to buy home.

Imagine 1 year loans to buy a pencil or pen or 100 years to buy a book, what kind of humanity will we be back then?

Imagine extra pay for those who have work and those who do not have work overhead payments and free food and transportation while they find them.

Imagine unlimited research and subsidies for paraplegic or disabled people or zero crime for limitless education and police with all means.

And democracies where rights are first of all in the splendor of tomorrow's humanity.

And thus seeing universal time

EF 33

FUTURE ECONOMI

How was globalization achieved?

And,how,would,that,economy,actually,be?

And this economy that we all want to know how it will be even with our personal economy, personal credits to 60 years, houses to be paid in 100 years of minimum payments, and fast personal approvals. The economy of the great future, where the automated society without limits for the human race will be achieved is detailed here, in the politics and the economy of the future in a world of government reserves where individuals will live better.

Set the foundations for that, and the projects about it that governments will study, and after some meetings that will be held in the high commissioners of the United Nations and the World Bank, they will get to a governmental agreement of a special currency, which will perhaps be called World Dollar, like the Euro dollars. This will create extra reserves for the support of all the currencies from all the countries by government order only, and in this way they will be substituting the established reserves, and everything will change, and then the world currencies of today will be benefit by an extra reserve in their countries for the needs they want. So, there will be an only currency. First, it will be issued for primary needs, society in disgrace like hunger and sickness, and then there will be great financings that will be substituting the old ways of gold warrantee, and backup of the currency of each country. Little by little and progressively until all of the currencies are totally substituted. The reserves of the countries will change in between 50 to 200 years, and then they will be so good that all the government and private projects will have financing and there will always be jobs for everybody, and budgets never seen before. There will never be limits for anything and in this way, little by little, everything will be substituted. By then, the normal reserves will be from the past, everything will be by a world and governmental decree, which will be capitalist, and there will be great partnerships that will be there to stay not only to create employments. But there will be no need of reserves for the currency, and the founds of the World Bank will benefit all the countries much better than nowadays, and our economies will come out of the almost Jurassic ways to which the territorial wars have gotten us into for the last 5000 years. And a thousand Years will finally come, and we will remember the ways of the barter from the past, even the Euro Dollar, and the ways that have only set back the politic division and have brought the lack of an homogeneous, politically civilized world, And that in the end of the times having nowhere else to go has gotten to its end. And so all the nations will be benefit for everybody, in an organized way until reaching a control range where we will have everything and where we will live in the famous leisure society of the future full of benefits with robots and androids that will do everything. The one that doesn't work won't need anything and will live in cities of culture only, without having deficiencies. That is the economy of the future... imagine sidereal journeys without financial limits, and journeys to the internal planets of our solar system with no limits, luxury hospitals for everyone and unlimited wonders, endless science investigations forever and no financial limit for medicines and culture in general, education and housing for everybody with this economy of the future...of the Years 2050 to 2200 to the future forever.

WORLD DOLLARS

On the other hand, the economy of the future will protect our patrimony and our reserves in an extraordinary way and as we have been saying, the patrimony of reserves of each country will be very secure. It will be a protection never seen before in history since in 5000 years this has never been done.

By using the special reserves, as the normal ones will be a backup only, with the time, between 30 to 100 years, they will be reserves only. Then, the official backup government reserves will substitute everything, and the economy of each country will be free of being pushed to have less value, or they will just not be affected when using the famous World Dollars of tomorrow or the government reserves, as you may call them (meaning World dollars or free cash reserves).

In this economy almost everything will be free, only that there will always be a state regulation for everything to be used. Even though the advantages of today, as we have said, are minimal and Jurassic compared with this economy so near to us, and that sooner or later will substitute everything established in the annals of the countries..And even though they will be governed, and only the countries that show more cleanness in their handlings will have more, there will be world regulator organisms. Maybe in the beginning the changes will show rapidly in all the countries in the world with great benefits for the humanity and its population, for the reserves of each country will not be affected as we have said, and the normal reserves will stay as basis for the economy only. Little by little, this huge was the total global unification with good imagination and the special education that was achieved for Washington as the unification of the planet would be acheved, called Global World of Today.

Future Economy 3

In the economy of the future, as we have said before, there won't be any deficiency or poverty because the governments will absorb all the economic waste of our society when possible, to be accurate, without affecting the emerging economies or the established private economies of the system itself as they are the poles that guarantee the employment, and also because of our democratic laws; but little by little they will substitute the special reserves to the budget deficit giving us all an investment field, but as described here, only after the total global unification or globalization this could be done. In the past, the countries' division would have never accepted that goal to be achieved, as stated and explained previously here, because they were at war. Here is even more detailed, and that is why we explain that by achieving globalization for all the reasons stated above, since all the countries divisions, speaking worldwide, prevented it from happening and made it always just some people's dream, and in other systems it could only be achieved with a total word invasion after purifying it for many years, decades perhaps, or centuries because they would slow down by their one general economic pole methods. Therefore, they could only be conceived in other methods like capitalism or what succeeded to it, ultra capitalism. And when at world peace, if it ever really existed, it was never real and only in world agreement and this total global unification world it will come little by little since, looking back in studies, before world peace is altered again it is known that nothing better could be achieved than in democracy, because in such a territorial world, as the earth of before, that was the result. It would only be achieved by substituting almost all the government, economically speaking, doing this in reality little by little aiming the wonderful tomorrow of humanity in years, decades an centuries, substituting all deficit of any country, company, individual, organization or charity group for them all to keep working as long as they are good and guarantee kindness, employment, supplies, and the essentials for everyone. And all the dreams of the perfect competition of the economy theories from the great economy authors of the past would be real someday. Maybe they would be kind of franchises and more, and they also will guarantee productivity and some rules that there will be, as long as there is an excuse for that, on the contrary of the by then qualified as retrograde systems of yore that felt threatened and couldn't do this because the world was divided worldwide and no one accepted other's wealth, only in equally situations of gold, oil, or normal reserves explained before and in universal literature. On the other hand, those skeptics were not the ones to blame but the war itself, and the world division, and what was alluded for years and millenniums since the invention of the reserve in the Roman or European eras, and later in the Egyptian eras of barter, and the Greco-Roman empire that migrated to Europe in the end and created economic division that came from the territorial division itself from the civilizations achieved in the past and that established a barter that would last 5000 years perhaps, as the war itself from Asia to America.

So, therefore, the skeptic scholar economists see that like a dream, if they even visualized it, if they even got to visualize it, because we are the first to suggest and write about it.We know this would be the only thing that would change or will change the world of tomorrow and today, besides globalization by implementing this economy that has held back humanity so much by not doing it and creating it and executing it in reality for the division and lack of the necessary proper parameters that couldn't be done without globalization.We just know that those scholars of yore, in their most intimate thoughts, were also allies of this possible economy, even though they visualized it in communist worlds or in commercial exchange of reserves, not like Juan C. Robles, this book's author. They never though if it this way, and we invented it when we presented it in our first edition of x99, because humans and human nature cannot be cheated, let alone history itself. They and the harshest were also allied in something similar that will generate in this someday, if my readers like my economic theories of tomorrow, which I think, little by little, they will think of as the only remedy to so much tragedy of humanity and good for tomorrow because they created other theories, but ours is more secure and better, but it could only be conceived in total globalization and that in reality, it will be this way or maybe very similar, because they thought of something like this, an economy that was not prejudicial in a complicated world from the beginning. And we know they will ally to this sooner or later because they won't find another better than this reserves of decree, and had they studied or study the global geopolitics, they would understand that the funds for this are given from an economy that won't hurt anybody, and so from bankers, giant companies and governments, to everybody is interested in that someone pays their unplayable liabilities, and in that everybody has something extra to buy. Therefore, in these days, in the beginning of the 21st century we know that with the democracies being unified, together with some factors of today, are the most appropriate time to do it, and so accept the economy of the future aided by the reserves of decree, and a world dollar that will substitute the normal reserves in decades or maybe centuries along with the most important governments of this world. And we know their descendants will embrace this system sooner or later, together with the modern democracies of the 21st and 22nd centuries, and that's why they will achieve this in 10 to 20 years or more since this wonderful global system that was accomplished has no better competition in a two to one bet in the wonderful economy of tomorrow in which there will circulating, issued in the shape of world dollars that will substitute debts and unaffordable liabilities, and the misfortunes of centuries will finally end, and so being protected by reserves of world government decree, as we explained before, they and those economies will prevail to get ahead in the future world full of surprises for everyone. Like those personal credits to buy a car in 30 years, or government personal credits to 100 years, credits for companies according to their needs to 100 years just for them to keep existing, and funds to banks to 200 years, which will really take us to the famous golden age of humanity, and to the great dream of the perfect competition from the economic theories written before competitiveness that in the past could only be possible in theories, will be a reality. Theories like the one of Adam Smith and the fathers of modern economy could be a reality.Then in these world economies, not impossible anymore Thank God but distant, a better world is visualized, or at least, there would be a remedy for the countries' world debts problems, private companies' debts, personal debts as well as for world hunger, and for unlimited medical research and space dreams of the past being true, and the ambitious conquer of outer space for free. And so, as the statistically perfect societies, or as perfect as possible, might be conceived, without supply or funding problems, in the happy worlds of tomorrow and their golden ages, and as was explained before and will be explained more in detail, and as conceived in reality because that's how it will be. Countries will get to agreements, previously explained, of reserves of government decree, and the global debts that seem to be hopeless will be the first ones to be protected, and the financing of the places that must be protected, so later, ensure employment and the guarantee of quality of life will be safe for everyone first. General services, like banks and private institutions, providing necessary services for everyone reaching homogeneity with more difficult problems. And so, little by little, the kindness of tomorrow will be admitted and then things will get better and better, and maybe the poorest nations in their hunger, doubts, and uncertainty will be the most benefited with this and also the richest nations in the beginning, in their super habits of research, social services and education; on the other hand, also the most powerful banks of the world in debt for unaffordable liabilities, and maybe debts to guarantee employment and everybody's economy; the giant companies that guarantee economy stability for the whole world and that were in danger of bankruptcy, and the economies' quality of life went from old to modern, from today to the future of extra reserves, where the requirement needed to obtain these kind of benefits will be at first perhaps, and later one day, maybe, as a funding right according to its services to benefit others.

All this in a global world order of approval from government sides.It will be wonderful that scientific achievements that may never had unlimited money were authorized with budgets for medical research, obtaining miraculous medicines and amazing scientific advance, which will give us a better, really civilized world each day. It is not necessary to point all this out, but this will be the only tomorrow awaiting everybody, statistically speaking.And by achieving this, it is observed that the society of the future has a good economy, there are no deficiencies, no calamities, they use an economy based in international reserves of government decree, and not in material reserves, but of decree, called World Dollar Agree or Free Cash; and not of gold, or oil or anything extra from the mentioned above, from what backed up the currency in the past, in that beautiful age of humanity, with unlimited budgets and unlimited medical, space and educational research, and social services. By that time, almost all the normal reserves that the countries contributed with for so many years, mentioned before, had been substituted and left only as funds, and this way, by unifying they mutually financed. And so, the economy of the future changed everything, and although they are still valuable for sale and part for the reserves for the future they may be substituted if appropriate, therefore there are no needs, and we'll be an almost golden society where there will be differences for our ancestor's jobs and condescending of better studies and competition, and society keeps being capitalist and democratic, with its rights of inheritances and big transnational companies because they also benefit of this, and their commercial emporiums or conglomerates are not overlooked, but there are no deficiencies, and the poorest have habitat, food, medicines and transportation almost for free; all this paid by the government and external subsidy; and everybody works for extra payment and for a place of, it could be said, wealthy middle class, and the boldest accomplish big companies, others, according to their skills, medium companies, and others external subsidy as previously mentioned. Everything is paid by the government, if worth it, with extra credit to a long term, in the end, there are banks for everything, and personal credits for cars to 40 years, and for houses to 60 years, ergo, the economy of the future absorbs everything on every space even the countries' debts are absorbed and swallowed by this wonderful plan of tomorrow and the foundation of an advanced society.

There are also conquers impossible to be made with normal reserves, like sidereal journeys and space advances that will be a fact not just attempts, and the recreation areas in the solar system like hotels in Mars and in other planets' orbits, not to mention galactic research, and one day, definitely, communication with beings from other worlds. Technology is huge; there are machines more intelligent than men, machines capable of copying automatically every object, atoms transmutation, materialization, almost instant teleportation and dimensional journeys, wireless energy transportation and fast trains that will avoid space time and will arrive in minutes to other cities like underground trains or floating in the open air in big magnetic shells at high speed; and they will get from Paris to New York, Los Angeles and Tokyo in minutes, and even to Mars; from one satellite city to another. In the South Pole, in cities that will grow as surrounding cities for research and that will grow along with the research stations of today as the famous Scotch station and others from America and international, because the government will spare no expenses for research and expansion of human life to places assimilating the life in external planets to migrate human life there, and so guarantee our species in tomorrow's space exploration odysseys, and life's conditioning in external planets, fabrication of elements based in control information of time, conditioning of jungles and grasslands in arid zones, weather control, mining in planets, automatic elaboration of every chemical product, exploitation of submarine products, directed rain, space energy, agriculture in the ocean, nuclear fusion, resistant plastic houses and new food for photosynthesis exploitation, electricity storage in super drivers, batteries, antimatter and antimatter reactors of unlimited, clean, ecologic energy, teaching machines, inanimate manpower, robots that will do everything, intelligent robots, communication with beings of other universe, and in general the biggest odyssey of the human being: emigration to other stars and the conquer of outer space on its first steps. There will also be machines to cure men by just sitting in a chair, and aging retardation, so men will stay alive 150 years that will seem 600 years, and changing other types of brains, bodies or machines as androids, and perhaps machines lengthening life enormously. That society of tomorrow will be wonderful; there won't be sickness or age, or old age; there will be incredible journeys through the solar system. Then, the whole world will become small; the cultured leisured society mentioned before will be accomplished, where buttons are just pushed, and work in many areas will be pleasurable, and so the extra wealth obtained will be enjoyed. There will also be a gamma of companies financed by the government with external subsidy that with very few excuses will keep working for years and decades, even if they are no profitable, just to provide price, supply and employment. This is why the economy of the future is competitive, giant, and very different from the ones in past centuries, therefore there are a thousand things financed by different governments like world banks of that time and thousands of types of different things, medicines and unlimited research as well as personal space journeys and a lot of government help that the today's individual doesn't have and that fills the guarantees of our societies of the future; in that time it will be normal that there won't be poor people, and the real golden age of humanity will be real. We will see how this economy takes us to a real humanity splendor that will be characterized by the lack of deficiencies, and by only working a little everybody will gain something extra, and the others could be in more continuous spreading than before in the so called, as we have said before, the leisure society of tomorrow and its magnetic trains through hexagonal gates of steel and alloys of cold function made instantly, and its incredible edifications that will dominate the elements, and the dimensional gates at high speed riding the planetary horizon interconnecting cities of the future, protected by magnetic fields, and wonderful underground desertic meadows with directed rain maybe from the 22nd and 23rd century views, and with starbases in orbit from interior planets like Neptune and Saturn and from this solar system of the 25 and 27, and journeys to other neighboring stars, and journeys to other universes, times, and spaces in the real golden age of humanity accomplished with the total global unification of the past, when we moved from a politically stagnant world to another more civilized that will take us, sooner or later, to that desired and planned economy of the future great for humanity............ THE END

-----------------------------------OO-------------------------------------

Thank you for your time

Sincerely yours

JUAN CARLOS ROBLES GUERRA

Realtradeiyerfci@gmail.com

011 52 662 563354

maul z cubillas 34 d hermosillo sonora mexico 80000

DIRECTORY

REAL TIME

THE LOCATION OF TIME IN THE FUTURE AND PAST UNIVERSE AND DIMENSIONS

CHAPTER 1

EF1 ETA CARIANAE POINT A1 A ETA C A64 IN THE REAL UNIVERSAL TIME EXPAND TIME

EF2 AUTHOR THEORIES ABOUT PROBABLE UNIVERSE

EF3 Time, as all subjects about the universe and space and specially time

EF4 DIMENSIONALS DENOMINATIONS -- IC1101 A1 TO 64--100

EF5 Alternate and altered lines of time do exist but a few times maybe artificially

EF6 AUTHOR THEORIES OF ATOMIC FORCE FIELDS BORDERING PARALLEL UNIVERSES

CHAPTER 2

EF7 AUTHOR'S ONRANGE THEORY TO BE PROVEN SCIENTIFICALLY

EF8 TIME SPLASH

EF9 That's why we expose, the universe was made from

EF10 REAL TIMELINE, IT HAS HAPPENED ONCE AND A THOUSAND TIMES AS WELL.

EF11 How many timelines could be created if we go to the past

EF12 TIME ESPACE CHANGE UNIVERSE TIME --BING BANG A1 ETACARIANE REFERENCE

CHAPTER 3

EF13 RJ LAWS

EF14 REAL TIME PART 2

EF15 On the other hand, many will ask, maybe, although in astronomy

EF16 CHAPTER 2 EF 2 SPACETIME MODERN ASTRONOMY REAL TIME 2

EF16.2 TEMPORARY RECORDING 2

CHAPTER 4

EF17 Back to the astronomy studies from 8 decades of the best astronomers

EF18 SPACE EMPTYNESS AND UNVERSE GRAVITY, HELIUM AND HYDROGEN EXPANDED AND ALMOST DISINTEGRATED.

EF19 them and the expanded atoms that must have something

EF20 Dimensional antimatter and steps to other universes and dimensions.

CHAPTER 5

EF21 H2O SPACE WATER

EF22 BLACK STARS, COMPRESSION HOLES OF ENDLESS MATTER.

EF23 UNIVERSES, WHAT ARE THEY?

EF24 ICAB4 TEMPORARY RECORDING

EF25 FUTURE PAST AND THE TEMPORARY RECORDING AND THENS OF TIME

CHAPTER 6

EF26 THE BLADES OF TIME AND THE UNIVERSE

EF27 The Orange Segments   As we know and believe faithfully

EF28 PRESENT PERFECT AND VARIANTS PRESENT INITIAL PERFECT

EF29 THE THINGS AND TEMPORARY RECORDS OF THE SUBJECT OF THE UNIVERSE 3

CHAPTER 7

EF30 THE GREAT ATTRACTOR AND OUR CHALLENGES

EPILG

EF31 THE RADIO OF TOMORROW

EF32 FUTURE ECONOMI PART 3.3

EF33 FUTURE ECONOMI How was globalization achieved?

-----------------------------------OO-------------------------------------

The architecture of tomorrow and
the conquest of outer space

Graphs of denominations past future time space

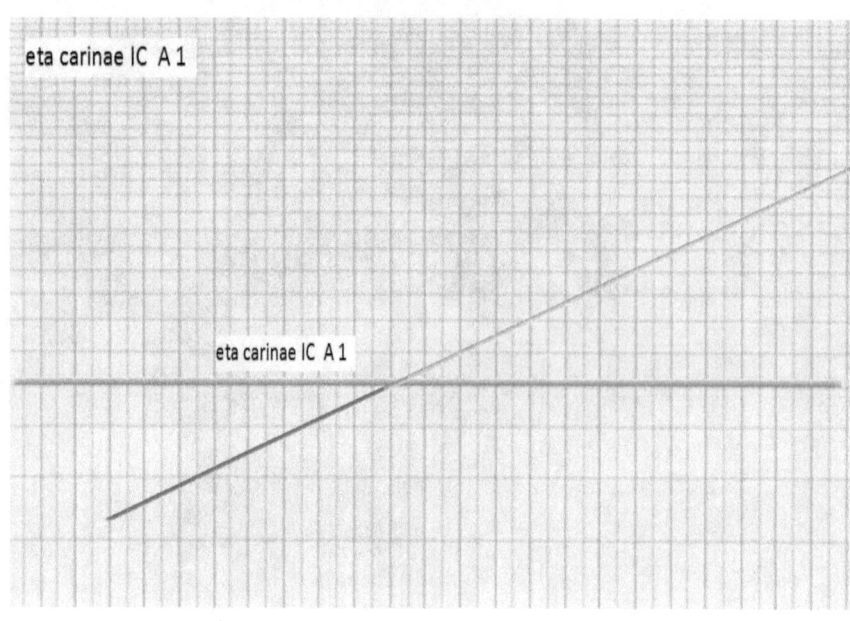

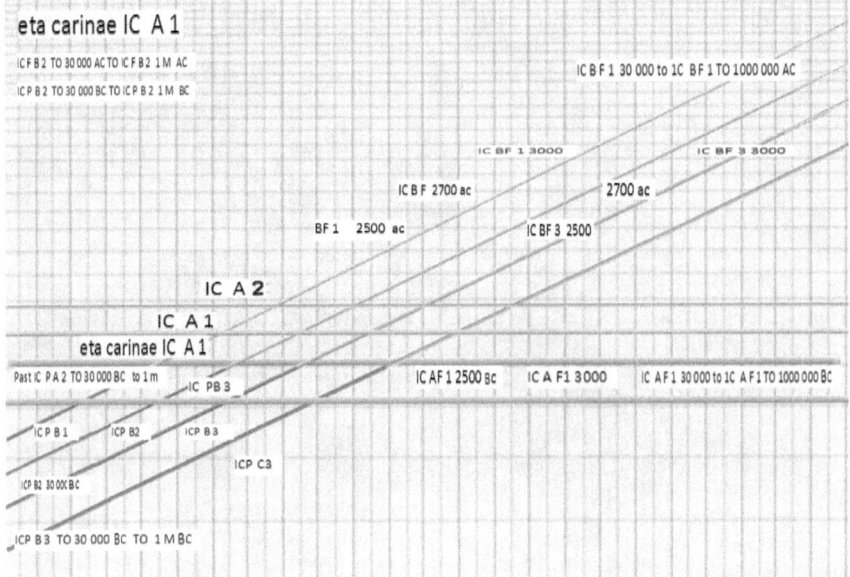

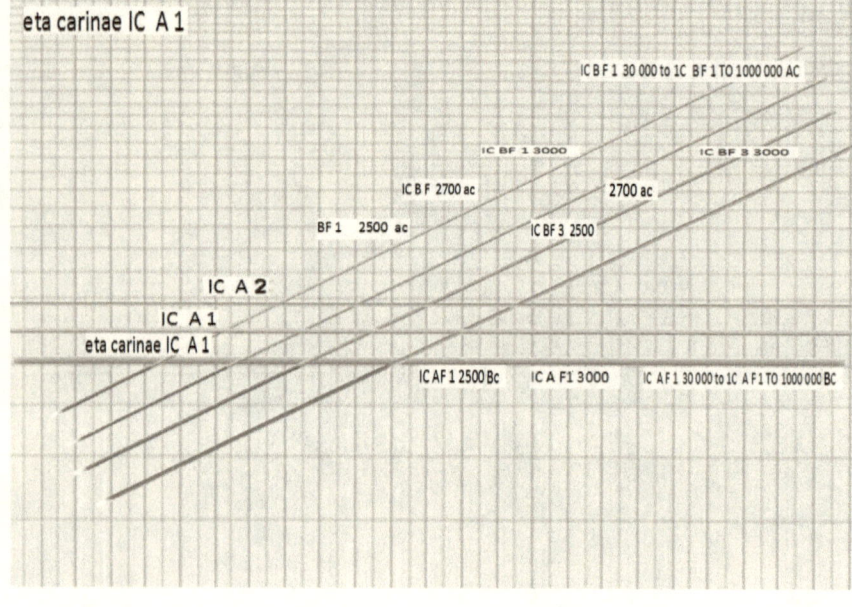

www.ingramcontent.com/pod-product-compliance
Lightning Source LLC
Chambersburg PA
CBHW021423170526
45164CB00001B/69